FARMER FIRST

FARMER FIRST
Farmer Innovation
and Agricultural Research

edited by
ROBERT CHAMBERS, ARNOLD PACEY and LORI ANN THRUPP

INTERMEDIATE TECHNOLOGY PUBLICATIONS 1989

Intermediate Technology Publications
103–105 Southampton Row, London WC1B 4HH, UK

© Intermediate Technology Publications 1989

ISBN UK hardback   1 85339 008 9
ISBN UK paperback   1 85339 007 0
ISBN USA hardback   0–942850–16–5
ISBN USA paperback 0–942850–20–3

Typeset by J&L Composition Ltd, Filey, North Yorkshire
Printed in Great Britain by Short Run Press, Exeter

Contents

ix

List of figures

List of tables

Preface

This book is addressed to all who are concerned with agricultural research, extension and development, regardless of discipline, profession or organization. It is for physical, biological and social scientists – agricultural engineers, agronomists, animal scientists, economists, entomologists, foresters, social anthropologists, sociologists, soils scientists, and many others; for researchers, extension workers, teachers and trainers; for those who work in International Agricultural Research Centres, in National Agricultural Research Systems, in Departments of Agricultural Extension, in Agricultural Universities, Faculties and Institutes, and in farmers' and other non-government organizations; for students who seek careers in agriculture; and for all – administrators, planners, staff of aid agencies, and NGO workers wherever they are, who are responsible for policy, management, teaching and training for agricultural research and extension.

The audience is wide because the topic is basic and the content challenging. Resource-poor farming in the Third World presents intractable problems. Probably well over a billion people depend for their livelihoods on the complex, diverse and risky forms of agriculture which have been poorly served by agricultural research. This failure has been met with two responses: 'more of the same' through the conventional generation and transfer of technology; and the development of new approaches and methods in which farmers participate.

In July 1987 some 50 people, with natural and social scientists in roughly equal numbers, met for five days at the Institute of Development Studies at the University of Sussex, UK, for a workshop on 'Farmers and Agricultural Research: Complementary Methods'. Many of those who took part had been developing new participatory research methods, some of them in isolation. We found that new and similar modes of agricultural research and development were being evolved in parallel in different parts of the world, but that most of the professional pioneers were in a minority and marginal in their institutions.

The term 'complementary methods' was used to avoid the impression that the new approach was an alternative or complete substitute for traditional on-station and in-laboratory agricultural research. The importance of commodity research and of farming systems research was acknowledged. There was concern, too, about the dangers of a new instant orthodoxy. At the same time, the new research methods appeared powerful and accurate in meeting farmers' priorities. The evidence presented indicated that these new approaches and methods could serve the complex, diverse and risk-prone agriculture which supports perhaps a quarter of humankind, with lessons also for all agriculture. We found we were dealing with a new paradigm, in the sense of mutually supporting concepts, values, methods and action. To this the term 'farmer first' has been applied, distinguishing it from the conventional paradigm of 'transfer of technology'.

The nature of the subject has demanded that we make this book more than a compilation of papers. In doing this, we have been helped by the preparatory research and analysis for the workshop carried out by John Farrington and Adrienne Martin and their paper 'Farmer Participatory Research: A Review of Concepts and Practices', which was revised and republished in 1988 as ODI Agricultural Administration Unit Occasional Paper 9. We are grateful to John Farrington, Janice Jiggins and others who have made comments on the text, and to the authors of the 42 workshop papers for their tolerance and understanding. We owe much to those who participated in the workshop discussions, some of whose verbal contributions we have tried to capture in the sections by 'IDS Workshop', which also include some of our own comments.

The brief summary preceding each of the four parts of the book is intended to help the reader gain a quick overview in a matter of minutes.

The workshop and the editing of this book were made possible through grants from the Ford Foundation, the Rockefeller Foundation, and SAREC. In addition, ISNAR enabled some participants to take part. A further grant from the Ford Foundation to the IDS has supported follow-up workshops in Peru and the Philippines, and other research and dissemination, especially by the Overseas Development Institute, London. The grants have helped this book to be published at a sustainably low price which we hope will make it accessible to all who can use it, especially those who work on low salaries and in countries with foreign exchange and recurrent budget constraints. The papers of a further follow-up workshop, convened by the Information Centre for Low External Input Agriculture, Leusden, Netherlands, are available (see appendix) as *Proceedings of the ILEIA Workshop Participatory Technology Development in Sustainable Agriculture, April 1988.*

We wish to thank Rhona Adams and Helen McLaren who organized the workshop. We are grateful to all those in many institutions who typed papers for the workshop and this book, and especially Helen McLaren in IDS who has throughout been the central point, calmly and competently handling the tasks of typing and managing a complex, diverse and risk-prone manuscript.

Permission from Cambridge University Press is acknowledged to include parts of six papers (by Baker et al, Kean, Lightfoot et al, Maurya et al, Norman et al, and Sumberg and Okali) which were published in a special issue of *Experimental Agriculture*, vol 24 part 3, 1987, edited by John Farrington. Also, the papers by Ashby et al, by Box, and by Rhoades, were distributed in December 1987 as Discussion Papers of the ODI Agricultural Administration (Research and Extension) Network. All these sources are listed in the references.

In the book we use the term 'farmer' as in the title. This is shorthand for the farm family, with special stress on the poorer and those with few resources, and on women, who are often and so easily neglected or left out. Many farmers are women. Often both women and men farm. The

xiv

importance of eliciting women's views, and of their playing a full part in the activities described in this book was a major theme in the IDS workshop and deserves repeated emphasis.

This book is not a final statement, but part of a process. It presents an outline of approaches, with evidence and examples. We have edited and written it to be convenient for teachers as a textbook for universities and institutes. We hope it will stimulate and encourage readers, of whatever profession or discipline, to learn from farmers' innovations, to put farmers' agendas first, and to support practical participation by farmers. Above all, we hope it will encourage many more to join in pioneering and writing, adding to and sharing experience and methods. For it is through hands-on experience and efforts to communicate that the practical potentials of farmer-first approaches and methods will spread and be realized.

The editors

Three types of agriculture summarized

	Industrial	Green Revolution	Third/'CDR'
Main locations	Industrialized countries and specialized enclaves in the Third World	Irrigated and stable rainfall, high potential areas in the Third World	Rainfed areas, hinterlands, most of sub-Saharan Africa, etc
Main climatic zone	Temperate	Tropical	Tropical
Major type of farmer	Highly capitalized family farms and plantations	Large and small farmers	Small and poor farm households
Use of purchased inputs	Very high	High	Low
Farming system, relatively	Simple	Simple	Complex
Environmental diversity, relatively	Uniform	Uniform	Diverse
Production stability	Moderate risk	Moderate risk	High risk
Current production as percentage of sustainable production	Far too high	Near the limit	Low
Priority for production	Reduce production	Maintain production	Raise production

CDR: complex, diverse and risk-prone

Introduction

The 1980s have seen shifts of thinking and priority in agriculture in much of the Third World. It has been increasingly recognized that questions about who produces food, who can command it, and where production takes place, often matter more than how much is produced. It has become clear that resource-poor families and conditions have been less well served by agricultural research than have resource-rich farmers. For reasons of both production and equity, rainfed agriculture has risen in importance compared with irrigated agriculture. Sustainability of output now also has a high place on the agricultural agenda because of widespread deforestation and environmental degradation. At the same time, population projections indicate that in many countries rural areas will have to support much larger populations, with many more people living in fragile and difficult environments. The priority has become not just sustainable agriculture, but sustainable livelihoods based on agriculture, not only for present populations but for hundreds of millions more people.

The thesis underlying this book is that these changes present a new challenge not just to agricultural policy, but also to agricultural professions. It is true that for adequate and decent livelihoods that are sustainable, much depends on policies which affect agriculture. An agricultural price policy based on paying good prices to producers is vital for a good or better living for rural people. Security of tenure and rights regarding land, water, livestock and trees are also preconditions for farmers to take the long view and invest in good husbandry, in trees, terracing and other physical works. Service infrastructure in roads, credit and input supply are also often important. But beyond these more familiar challenges, lies one that is deeper and less obvious. This concerns what can be termed normal professionalism – the thinking, values, methods and behaviour dominant in a profession. The thesis is that for the new priorities in agriculture, normal agricultural professionalism is part of the problem.

This can be understood in terms of the three types of agriculture identified by the Brundtland Commission (WCED 1987: 120–2) (see opposite). These were industrial agriculture, green revolution agriculture, and the third, resource-poor agriculture. The first or industrial agriculture is found mainly in the industrialized, rich world, but also in specialized enclaves in the Third World. It has large farming units, is highy capitalized, and relies on high inputs and often on subsidies. The second or green revolution agriculture is found in agricultural heartlands in well-endowed areas in the Third World, either irrigated or with good and reliable rainfall. These include the large irrigated plains and deltas of South, South-east and East Asia, and parts of Latin America and North Africa. It includes large and small farms, and exploits high-yielding varieties with complementary inputs.

The third type of agriculture has been variously described as 'low-resource', 'resource-poor' or 'undervalued-resource', and is identified with unfavourable or difficult areas. These are mainly rain-fed, and often undulating and with fragile or problem soils. They include farming lands of many types – in hinterlands, high lands, drylands, and wetlands, and in forests, mountains and hills, savannas, near-deserts, and swamps. Examples are the Deccan Plateau in India, the uplands of many countries in Southeast Asia and Latin America, and most of sub-Saharan Africa. According to one estimate (Wolf 1986), some 1.4 billion people, or over a quarter of the human race, are dependent on this form of agriculture for their livelihoods, comprising approximately 1 billion in Asia, 300 million in sub-Saharan Africa, and 100 million in Latin America.

The new challenge to agricultural research can be understood in terms of these three types of agriculture. Industrial and green revolution agriculture are both relatively simple in their farming systems, often with large fields and monocropping, uniform in their environments, and low-risk. In contrast, the third agriculture can be characterized as complex in its farming systems, diverse in its environments, and risk-prone.

Several factors have contributed to the success of normal agricultural research with industrial and green revolution agriculture. One is that conditions on research stations, with controlled environments and easy access to inputs, have usually been close to those of resource-rich farmers: what does well on the research station can therefore do well, other things being equal, with the farmer. Another is that the standard methods of agronomic research have generated high input packages which are simple and amenable to widespread adoption in uniform and relatively low-risk environments. Yet another factor is that the sorts of farms and farm families best able to benefit – those which are resource-rich, with good farming conditions and good access to capital, inputs and markets – have been well represented in the main industrial and green revolution agricultural areas; and in green revolution areas, many smaller and poorer farm families have also gradually managed to make some gains from the new technologies.

In contrast, the resource-poor farm families of the third – complex, diverse and risk-prone – agriculture have not benefitted or have not benefited as much. In contrast with industrial and green revolution agriculture, the physical, social and economic conditions of this resource-poor agriculture differ more from those of research stations. Simple and high-input packages do not fit well with the small scale, complexity and diversity of their farming systems, nor with their poor access and risk-prone environments. For them, as Paul Richards (pp 39–42) describes, each season demands its own adaptive performance, depending on unpredictable weather, and the interplay over time of farming activities with the household's resources. Farm families often lack reliable access to purchased inputs, and need to use them sparingly, if at all, in the face of risks. In these conditions, there are limits to the extent their needs can be met by conventional research.

One consequence has been that resource-poor farmers have been slow or unable to adopt many of the recommendations flowing from agricultural research. In the 1950s and 1960s, non-adoption was often attributed to ignorance, and extension education was prescribed. In the 1970s and the earlier 1980s, non-adoption was more often attributed to farm-level constraints; gaps in yield between research station and farm were analysed; and the prescription was to try to make the farm more like the research station. In the 1980s, however, a new interpretation, more challenging to the agricultural professions and to science, has gathered support. It is that the problem is neither the farmer nor the farm, but the technology; and that the faults of the technology can be traced to the priorities and processes which generate it.

This insight has many sources: world-wide, indigenous technical knowledge has been more and more recognized as valid and useful; in agriculture, social and biological scientists have increasingly gone to farmers to understand reasons for non-adoption; farming systems research has made a huge contribution by revealing the complexity of farming systems and of the decisions which face resource-poor farmers, and the limitations of multi-disciplinary statistical analysis; farmers have increasingly been recognized as themselves innovators and experimenters (Johnson 1972; Richards 1985; Rhoades and Bebbington 1988); and perhaps most decisively, farmers have again and again been found to be rational and right in behaviour which at first seemed irrational and wrong to outside professional observers. While these changes have been gathering momentum, a small minority of social and biological scientists, and of fieldworkers in non-government organizations (NGOs), have been collaborating in new ways with farm families, and showing that besides normal agricultural research, there are also other ways to identify priorities and to develop and test technologies.

As so often happens in the early stages of a new movement, many flowers have bloomed and many labels have been used. 'Farmer-back-to-farmer' (Rhoades and Booth 1982), 'farmer-first-and-last' (Chambers and Ghildyal 1985), 'farmer participatory research' (Farrington and Martin 1987), and 'Approach Development' (Scheuermeier 1988) have been added to their precursor known sometimes as 'downstream' farming systems research (Gilbert et al 1980). The later forms of these approaches all use reversals to complement conventional research. The conventional approach has been 'transfer-of-technology'. In this mode, priorities are determined by scientists, who generate technology on research stations and in laboratories, to be transferred through extension services to farmers. In the new, complementary mode, this process is stood on its head. Instead of starting with the knowledge, problems, analysis and priorities of scientists, it starts with the knowledge, problems, analysis and priorities of farmers and farm families. Instead of the research station as the main locus of action, it is now the farm. Instead of the scientist as the central experimenter, it is now the farmer, whether woman or man, and other members of the farm family. The label that is given to these practices does not matter. But as contributions to this book show, farmers' participation and priorities are recurrent themes, and reversals too are central. Together, these elements can be described as 'farmer-first'.

Farmer-first approaches and methods constitute a complementary paradigm. 'Complementary' is used, since the transfer-of-technology approach, including commodity research, on-station and in-laboratory basic investigations, and so on, will always be needed. 'Paradigm' is used since it carries the sense of a mutually supporting pattern of concepts, analysis, methods and behaviour. The contributors to this book include biological and social scientists who have been leading in the exploration and development of this paradigm. Within their own organizations they have often been a minority, evolving views and methods which some still consider heresy. With their contributions brought together here, it is evident that they are working on similar lines and with similar good results. In the paradigm they explore and describe, farmers are primary: it is they who come first and who identify their own priorities; and it is they who are the key actors, choosing, experimenting, and adapting in order to survive and do better.

The ideas and evidence in this book are not finished and final. They are a stage in a process of pioneering and learning. They do, though, point towards solutions to the intractable problems with which we started: the need for the third agriculture and for resource-poor farmers to produce more and to generate many more sustainable livelihoods. They raise the question whether the potential of the third agriculture has been underestimated. For the low potential may be partly only apparent, an artefact of inappropriate technologies which do not fit. The issue now, and for the 1990s, is whether with farmer-first approaches, major gains in production, incomes and livelihoods can be achieved; and if so, how these approaches can be widely and rapidly developed, diffused and adopted in the agricultural professions.

Each of the four parts of the book concentrates on a main theme. The first concerns farmer innovation, with evidence of the capacity of farmers, especially resource-poor farmers, to experiment, adapt and innovate. The second part concerns methods to enable farmers' agendas to be put first. The third deals with practical participation by farmers. Finally, the fourth part considers the implications for institutions and action, and what is needed for the future.

PART 1

Farmer innovation

Introduction

The theme of Part 1 is that farmers, especially resource-poor farmers, continuously experiment, adapt and innovate. Robert Rhoades uses historical
and comtemporary evidence to show how farmers always have been innovators and how they still are. DM Maurya gives examples from rice cultivation
in India which show how farmers select their varieties and how their
criteria can differ from those of scientists. Dianne Rocheleau and her co-
authors argue that agroforestry demands the invention of research methods
based on ethnobotany and agroecology, including use of local knowledge,
chains of interviews and farmers' experiments on home gardens. Anil
Gupta shows how scientists can overcome barriers between themselves and
between themselves and farmers, leading to learning with and from each
other. While normal science generates packages, resource-poor farmers, as
Paul Richards explains, engage in farming as a continuous performance.
    Farmers' ability in this performance to classify, choose, improvise, adapt
and test is illustrated by examples from potato storage technology, seed
variety selection, agroforestry, tool-making, the invention of complex
cropping patterns, soil conservation, water harvesting, and uses of native
species. When farmers are seen in this light, as experimenters and
innovators, other views also change: what farmers need is less a standard
package of practices and more a basket of choices; the role of extension is
less to transfer technology and more to help farmers adapt; the local
experts are not so much researchers as farmers themselves. Farmers are
professional specialists in survival, but their skills and knowledge have yet
to be fully recognized. Thus one purpose of this part of the book is to show
how scientists have discovered or elicited from farmers ideas and information, techniques and knowledge. From this, one can begin to appreciate
the possibility of more dynamic and flexible research processes, building on
farmer-researcher interactions and supporting farmer innovation. These
processes are examined in Parts 2, 3 and 4.

1.1 The role of farmers in the creation of agricultural technology

ROBERT RHOADES

A legacy of farmer innovation

To many, the thrust of this book will be a heresy: that farmers have much
of importance to say to scientists and that farmers' methods of practical
research are complementary to those of scientists. We have long accepted

3

that scientists have something worthwhile to say and give to farmers and have advocated the transfer of technology from scientists to farmers: international agricultural research centres, national agricultural research systems and extension agencies are based on the experts' authority. This book does not turn that on its head to suggest that scientists have nothing worthwhile for farmers. Rather it posits that farmers' knowledge, inventiveness and experimentation have long been undervalued and that farmers and scientists can and should be partners in the real and full sense of that work in the research and extension process.

The thesis that farmers have an important role in agricultural research logically leads to two questions. First, what is the empirical, as opposed to romantic or emotional, basis for elevating farmers to an equal partnership in research development? Second, how do we match up the comparative advantages of each class of specialists (scientists and farmers) in a truly meaningful way?

Two kinds of evidence can be used to demonstrate the importance of matching farmers' concerns and innovative capabilities with scientific methods. One kind pertains to long-term contributions of farmers to modern agriculture; the other is specific recent innovations (or research methods) in which farmers have played a part.

I believe that the scepticism about farmers' knowledge and potential contribution stems from an honest appraisal on the part of many dedicated scientists. Researchers simply have not seen hard evidence to prove or disapprove its existence and value. This is partly because farmers seldom record their accomplishments in writing, rarely write papers on their discoveries and do not attach their names and patents to their inventions. As a result, the history of agriculture is written without reference to the main innovators in the long-term process of technological change. Moreover, academic disciplines which one might expect would have documented farmers' contributions, such as economics and anthropology, have not done so.

The archaeological and historical records, however, reveal a very different reality. Braidwood (1967) discusses the 'atmosphere of experimentation' which characterized the Neolithic farmer since the earliest stages of agriculture. Farmers selected and domesticated all the major and minor food crops on which humankind survives today. Early cultivators knew about the characteristics, food value and medicinal uses of over 1,500 plant species. Over 500 vegetables were cultivated in ancient times. Moreover, agriculture did not originate in just one or two centres. The best evidence on early domestication shows that experimentation with all the important semi-wild crops was occurring simultaneously in different areas of Asia, Africa, Europe and the Americas (Reed, 1977). Later many types of hand-tool and ultimately the plough were developed. As Johnson (1972:156) has argued, variation and experimentation are the 'basic stuff of which adaptation and evolutionary change are made'.

In pursuit of this theme, I conducted a futile literature search for publications on farmer-originated technologies. Many authors refer to the technologies themselves but do not make reference to their creators. For example, it is rare indeed to find any discussion of the suggestion that it was

4

women rather than male cultivators who domesticated many plants and invented grain milling (Mozans, 1983:343).[1] The historian Fussell (1965) is the only scholar I encountered who recognized the contribution of farmers:

The main achievement of farmers, helped by scientists and engineers, has been to produce crops which are more prolific and of better value in human nutrition and to breed animals that put on flesh where man finds it best for his requirements, and generate more milk, or grow better wool. In total, this is no mean distinction.

Further reasons why farmer innovation has not often been discussed are associated with the eternal optimism of the 1950s and 1960s about the benefits of western science and technology. During those decades, the 'green revolution' was pulling India from the brink of starvation. Throughout the Third World, food production was rising, and both the First and Second Worlds were facing bumper harvests. The basic assumption of development efforts was that a large backlog of scientific information and technologies were stockpiled ready to be 'transferred' from experiment stations to farmers.

I would like to rethink the 'transfer of technology' model of agricultural research and development by drawing on three studies which I have researched where farmers played a creative role in technology generation.

Diffused light technology: a farmer idea

Robert Booth, myself and others have received much professional credit for the work on diffused light storage of potatoes carried out at the International Potato Centre (CIP) in Peru. Few people realize, however, that this was a technology which CIP scientists first learned from Third World farmers. The story began in the early 1970s when Jim Bryan, a CIP seed specialist, first observed farmers in Kenya storing potatoes in diffused light. Later in Nepal and in Peru itself he saw other versions of the same technique. He persuaded colleagues to investigate these farmer practices and when they had tested and refined the principle and passed the idea back to farmers who had not used it before, diffused light storage became a rather well-known case of successful technology development by an international centre.

When Booth and I conducted a follow-up in several countries, we were surprised that adoption had not proceeded as we expected. Out of some 4,000 cases, at least 98 per cent of the farmers had not 'adopted' the technology as it was presented in extension efforts, but had 'adapted' the idea to their own farming conditions, household architecture and pocket books. Farmers do not think in terms of adoption or non-adoption as we do, but select elements from technological complexes to suit their constantly changing circumstances. The dichotomous terms of adoption, non-adoption, traditional-modern, native-improved, are irrelevant and misleading from the farmers' point of view. Farmers did not drop the old storage practices (seed kept in darkness) to adopt diffused light storage. Instead, the vast majority just incorporated

5

the diffused light method along with their existing practices. They frequently used only those elements of the diffused light storage packages which interested them. Thus farmers rarely built a 'model' potato store, but modified the principles of diffused light storage to their conditions, designing alterations to their existing store – a creative input they enjoyed tremendously.

Social factors were as important in the redesign process as technical or economic considerations and included prevention of theft, privacy and gender control. Moreover, most farmers began by experimenting with small quantities before moving to larger investments. Adoption was step-wise unless there were other reasons to build a larger, model store similar to that of the extension service (to get credit or rights to better seed from the government).

This experience taught us that farmers and scientists had indeed much to learn from each other, and particularly we had much to learn from farmers. Farmer-teaching-scientist does not rule out the need for scientific research. What it does do is open a new area of research: understanding farmer technologies and systems for the use of farmers in other geographical areas. Even now it seems that much technology promoted by development agencies may have been learned from farm people, directly or indirectly. An anthropologist at the International Rice Research Institute who compiled a list of technologies on offer from the Institute found that 90 per cent of those being promoted had been derived from Asian farmers (Goodell, 1982). The ideas for them had been brought to IRRI by Asian researchers who came for a year's sabbatical with the aim of leaving behind a workable technology. Most left behind an idea or technique that the farmers from their country had been using for generations.

The great germ-plasm issue

Farmer adoption of new varieties and the problem of 'genetic erosion' has received international attention during the past few years. Most of the debates, however, are based on dataless accusations rather than on hard evidence of what is really going on at the farm level with new varieties. Both sides in the controversy tend to see farmers as passive and somewhat unthinking individuals. Yet farmers have been dedicated plant and animal breeders for thousands of years, although not in the precise manner of modern genetics. They have consciously maintained diversity, planted mixed fields systematically to achieve natural crosses, practised selection and set up their own personal gene banks as well as far-flung exchange systems for acquiring new genetic material. By the same token, however, farmers are not so naive as to throw away their older varieties and production strategies simply because they are presented with a new package. Let us consider potatoes in developing countries, although rice, maize or beans would probably show the same.

Over the past five years, I have been leader of a project to examine farmer selection and use of both local and introduced varieties of potatoes. Based on a review of the literature, we expected to verify the following patterns:

6

- farmers were reducing the number of their varieties and dropping native landraces in favour of improved ones;
- improved varieties were grown exclusively for commercial ends, native varieties for home consumption;
- farmers valued improved varieties for yield and market qualities and local varieties for taste, storage and cooking/processing quality;
- wealthier farmers adopted hybrid varieties, while poorer farmers clung to traditional ones;
- improved varieties required pesticides and fertilizers; native varieties did not need these due to natural resistances.

When we went to the field, talked to farmers and tested these commonly accepted ideas, we were again surprised in much the same way as with diffused light storage. Farmers were doing things their way rather than according to scientists' preconceived ideas. We discovered that farmers do not draw the same distinction as scientists between 'improved' and 'traditional' varieties. Rather than improved varieties pushing out native varieties, we found farmers studiously incorporating the new material into their private germplasm banks which contained an average of six to seven varieties. We discovered that some communities grew native varieties for the market while poorer communities and families grew the higher-yielding improved varieties for home consumption because they produced more food per square metre. In these communities, farmers even considered the tastes of improved varieties equal to some native landraces. We found some improved varieties stored better than native varieties and some native varieties yielded more than improved ones. In other words, the 'hybrid-native' distinction was again that of outsiders. In fact, a new variety that was around for any amount of time could pass in status from 'improved' to a native variety in the minds of farmers.

We also learned that farmers incorporated germ-plasm materials into their banks in a logical manner. Farmers are fanatic seekers of new varieties as anyone who has ever worked in a seed programme can testify. Once a new variety is obtained, they begin by planting a few in a kitchen garden or a single short row along the boundary of a field. They watch and observe. If the variety proves itself, farmers amplify their production, restricted – of course – by the amount of available seed. Finally, they will put more and more of their land in this variety. All the while, they continue to maintain their own 'germ-plasm' bank which is constantly being replenished and culled. We may call the latter 'genetic erosion' but farmers do not perceive the dropping of varieties as a loss but rather as a decision as part of their farming operations.

Creating a new farming system: the pioneers of Tupac Amaru

The third case involves creation of a new farming system. Several international centres are devoted to the creation of totally new farming systems, an area where they feel they have a comparative advantage (Swindale, 1987), especially for ecologically marginal zones such as deserts or tropical

7

forests. In the yearly CGIAR meeting on farming systems research at ICRISAT in India, much scepticism was expressed about the role of farmers in developing new systems. This was seen as largely a research process which should take place on experiment stations, to be later transferred wholesale to given geographic areas. Such a complex undertaking was seen as beyond the capabilities of farmers.

For more than 15 years, scientists from many nations have been working in the jungle area near Yurimaguas, Peru, attempting to design a new farming system as an alternative to shifting cultivation which is low yielding and destructive. The research effort focuses on improved technologies and methods of soil and crop improvement to allow farmers to intensify on smaller areas of land and to practice cultivation which is intensive and permanent instead of shifting. Despite high standards of research, few farmers have yet taken up the more intensive system, preferring instead to maintain their shifting cultivation methods.

Directly across the river from the experiment station, a new farming community named Tupac Amaru has been emerging (Rhoades and Bidagaray 1987). The pioneer farmers came to Yurimaguas from the highlands and coast with the same goals as the scientist: to intensify production in the jungle area. Instead of looking at ways of improving the jungle soil and intensifying with local crops they brought with them knowledge of upland rice cultivation from the coast and upper montane zones. Wet rice (irrigated rice), with which the farmers had prior experience, had never been tried before in Peru's Amazon Basin.

In 1981, without help or prompting from the government, a pioneer group of these farmers visited Yurimaguas for the first time to select their new wet rice area. Since then some 120 families have come to Tupac Amaru.

The farmers of Tupac Amaru started off slowly by mixing traditional rice methods of production with irrigation. They built canals and experimented with new drainage systems. By 1984, they had 280 hectares under rice and planned for 600 hectares by 1985, shooting for 5,000 hectares in the future. Now that they have established themselves, several banks, government agencies and even international agencies have come to help (and also to claim credit for developing this revolutionary farming system).

Conclusions

The similarities between what farmers do and the scientific methods seems clear. In the storage work, farmers did not run out and build a new diffused light store. Instead, they took first a handful of tubers, placed them on a window sill and watched. After a few months, they compared shrinkage, sprout elongation and overall seed quality with tubers from the same batch which had been stored in darkness. In the variety adoption case, farmers planted small rows along the fields of their other varieties or in the household garden. They watched for resistances, measured the time to maturation, noted the colour, sized up the tubers and even cooked them for taste. Finally, in Tupac Amaru, there was constant experimentation, not only with irrigation and drainage systems and seeds, but with social arrangements to overcome credit, transportation, and subsistence problems. I also observed complex on-farm trials with replications designed and run by farmers without any technical advisers.

The chief difference between these examples and what scientists do is that farmers have very specific goals in mind and the results of experiments must be practical. Farmers have less room for investigation purely out of interest. In personal correspondence, my former colleague Robert Booth has summarized the position succinctly:

> It is my general experience that a vast majority of research workers prefer to do research about a problem rather than research to solve a problem. Thus, biological scientists keep busy, and happy, breeding new varieties, developing disease control systems, or new store designs, while the socio-economists undertake their surveys and describe systems but all leave the actual solving of farmers' (clients') problems to someone else, and hence we hear of poor extension services and backward farmers. This, to my mind, is simply passing the buck . . .

During the 1970s, an important catch-phrase in agricultural development was the 'yield gap'. The problem was clear-cut: why were farmers not capable of obtaining yields on their own farms comparable to the high yields on the scientists' experiment stations? This led to 'constraints' research, which in turn led to farming systems research. Our objective here is to carry the discussion even further. We now realize that while the experiment station is the right place to conduct basic research (eg, asking how do potatoes grow?), it has limitations for real-life applied research (eg, how to grow potatoes?). The next 'gap' that we need to close is the farmer-scientist gap, or we will remain in the situation one Nigerian described in his own country: 'The scientist is as distant to the farmer, who the scientist claims to be benefiting by his research, as the moon is from the earth' (Alao, 1980).

1.2   The innovative approach of Indian farmers

D M MAURYA

Selection of crop varieties[2]

The green revolution vastly increased the productivity of some major crops in India, but it left many of the problems of resource-poor farmers untouched. Part of the reason is that most research stations are situated on ideal land and do not represent stress situations. Technology which works well on the research station may prove to be of no advantage to farmers on less well-situated land. One response has been the development of on-farm research, in which we offer farmers technologies with alternatives and look for the most appropriate approach in conjunction with the farmers. However, on-farm research should never be regarded as a substitute for experiment station research as many forms of basic investigation can only be undertaken on station.

My experience of this is more fully discussed elsewhere (Maurya, Bottrall and Farrington, 1988). For the moment I wish to stress that even in the absence of on-farm research involving outsiders, farmers regularly innovate and make their own selection of appropriate technologies. Sometimes, indeed, they select technologies which have been rejected by official research. The most striking example is the paddy variety Mahsuri which was introduced into India from Malaysia for tests during 1967–68. After two years of work, this variety was rejected by rice breeders on account of its lodging behaviour. But somehow the seed reached some villages through a farm labourer in Andhra Pradesh. Farmers who tried it found its performance excellent. As a result, it spread from Andhra to Orissa, and then to West Bengal, Bihar, Uttar Pradesh and part of Madhya Pradesh (figure 1.1). As a result of this 'farmer-to-farmer' extension, Mahsuri is now the third most popular variety among Indian farmers, after IR8 and Jaya dwarf rice. Its semi-tall habit, high tillering, heavy panicle, high milling out-turn and excellent grain quality make it well-liked by farmers. For quite some time the variety was not officially released by government, but because of its popularity, demand for quality seed and pressure from farmers, the Government of India was forced to notify the variety under the Seeds Act so that, duly certified, demand for the seed could be met.

A similar instance occurred in Uttar Pradesh when a breeding line, IR 24, which was under test, 'fled' in the hands of a farmer who began to grow it. Being susceptible to low temperature under late planting conditions, most of the plants did not flower. The few that did were harvested by the farmer and grown next year. This selected plant population performed excellently. The farmer multiplied the seed and other farmers grew it. Eventually the farmer wrote a letter to the Uttar Pradesh Government asking for its official release under the name of Indrasan. Meanwhile, the variety (or line) spread further and crossed state boundaries into Haryana. When an epidemic of white-backed plant hopper occurred in Haryana during 1985, this was the only rice which stood in the field safely. After such experience of its pest resistance, its popularity increased many-fold.

There have been similar experiences with another rice variety, Sarjoo 49, which continues to be widely grown despite being officially withdrawn and the same sort of thing has happened with a sugar cane variety, B.O. 17.

When hybrid maize was first released in India, its performance was often disappointing. The reason given officially was its susceptibility to erratic irrigation, especially at the reproductive stage. However, there was a social reason also. For consumers using green cobs for roasting, the hybrid cobs were bigger than they needed and they were reluctant to pay extra for them. Thus farmers reverted to local types with medium-sized cobs and moderate resistance to water stress.

10

Figure 1.1: *States within India mentioned in section 1.2, illustrating by arrows the spread of paddy variety Mahsuri by 'farmer-to-farmer' extension.*

In North Bihar and some eastern Uttar Pradesh districts, the new maize caught on, not in the season which researchers had expected but in the rabi (winter) season. There is a tradition in Bihar of growing rabi maize and when hybrid maize was grown along with local varieties, farmers were impressed.

One other example of farmers' astuteness in selecting varieties refers to an American cotton grown in north-west India. This did not fit the local cropping system because of its late maturity. However, when growing the American cotton, one farmer from Bikaner, Rajasthan, selected an early plant, multiplied the seed and started growing it as a distinct variety. This fitted well with the cropping system and the yield and quality were good. Within a short period, the new variety became very popular under the name Bikaneri Lerma. It is now grown throughout the cotton belt of Rajasthan, Punjab and Haryana. For many years since then, professional cotton breeders have been engaged in searching out defects in this variety, wanting to pull it down. However, its popularity prevailed and the voices of farmers forced its official recognition. This is a fine example of a farmer working as a breeder without any school or college training in genetics.

Machines, pest control and fertilizer

After the introduction of dwarf wheat during the green revolution, new technologies were introduced in the form of wheat threshers, pumpsets and comb plankers. All these underwent various sorts of modification at the hands of local artisans to suit small and medium farmers. Several local types in small sizes were introduced and perfected by artisans and have been widely used, even by small farmers.

Regarding weed problems, there has been particular difficulty with *Phalaris minor* in wheat and *Echinochloa* species in rice. Weed scientists in Punjab and elsewhere recommended spraying with herbicide solutions. Some innovative farmers who lacked a sprayer mixed the recommended herbicides with sand, or even urea, and spread them broadcast. This worked well and the technique spread like wildfire. Many queries were addressed to the scientists concerning the validity of this procedure. After testing at the experiment stations, the technique was found to be nearly as effective as using a spray.

A traditional system of weed control in India about which many rice scientists (including myself) were sceptical is the Beusani system. Fields in which a 25 to 35-day-old rice crop is growing are ploughed crosswise very gently, with about 5 cm of standing water in the fields, and planked. This serves the triple purpose of weeding, thinning and interculturing the crop. The major weed of rice, the *Echinochloa* species, is similar to the rice plant in its early stages but develops internodes within 30 days. During planking, most of the weed shoots are broken and detached at their nodes whereas the rice plants do not develop such nodes and thus escape breaking losses. During planking, rice plants are simply bent down and after some time resume their original erect position. The farmers with their keen observation and innovative approach could identify this differential behaviour in

12

the growth habit of rice and rice-like weed flora. Unfortunately, this most scientific technique was not appreciated by rice scientists and no effort was made for a long time to improve it. At the Narendra Dev Agricultural University, when we started to work in the villages and saw the technique, we were very impressed by the explanation the farmers gave. This was a turning point in making us realize that local practices are not altogether irrelevant.

A more recent innovation comes from eastern Uttar Pradesh where farmers noticed a severe abnormality in the paddy crop which responded to the pyrites being used for reclaiming saline/alkali soils. As a result, they approached me as a rice scientist, to ask whether application of pyrites to the crop made sense and, if so, what would be the proper 'dose'. No experiment had been done on this until the farmers approached us and we could only offer advice after conducting trials at the experiment station. The results of these trials were very encouraging and the use of pyrites is now an important component of fertilizers used in the area.

Farmers' practices not only demonstrate striking innovations, but often indicate important points overlooked by research workers. During one survey, when farmers were asked why they did not heed advice to plant crops in rows, they had to explain that when all tasks are done manually, much extra time and labour is required in using a rope to set out straight lines for planting.

Resource-poor farmers with small amounts of land also practise multi-cropping for the sake of insurance. Unfortunately, plant breeders have not taken into account the plant characteristics that favour growing crops together. The new varieties of crops are not well suited to mixed systems. Yet it has been estimated that companion planting of this sort can increase both yield and profitability by about 50 per cent.

Farmers as partners in varietal selection

In view of the difficulty of breeding appropriate rice varieties for resource-poor farmers in rainfed upland areas and of the skill in selection which farmers show, we are now working with them to test new varieties (for a fuller description of these procedures and their evaluation see Maurya et al 1988). A large number of advanced lines of crops with resistance to prevailing diseases and insects and a wide range of grain types, height and other traits are being investigated. The farmers are doing the testing and choosing varieties for themselves, not individually but jointly.

Whilst individual farmers are experimenting with just one or two breeding lines, after flowering and especially at maturity, all the farmers in the village assemble and visit the experiments together. The farmers are asked to state their choice based on visual field performance, subject to final opinions being given after harvesting, threshing, milling and cooking. A large number of advanced lines are tested in a single village and in neighbouring fields, so the farmers have access to all genotypes raised in their village from the earliest stages to maturity. During this period they can discuss the merits of each one amongst themselves. In the course of this

13

work, they have identified NDR 112 and 132 as good lines under their poor environmental conditions. On the experiment station, these varieties do not seem so good.

Procedures for testing have to be modified when farmers do the work. At experiment stations, all treatments are replicated to give valid estimates of error and better precision. However, in on-farm research with resource-poor farmers, it is very difficult to conduct replicated trials with many breeding lines. To overcome this problem, each genotype/line was given to at least three farmers, repeating the same set in at least three villages with approximately identical conditions.

Another procedure followed at experiment stations is randomization of treatments in the plots. In on-farm trials the breeding lines allocated to each farmer are not strictly random. Farmers are given lines comparable with the local variety they would otherwise be growing at least in terms of maturity and grain type.

The other peculiarity of the new approach is that rather than provide a package of management techniques and materials, the farmers are free to practice the same level of management in the trials as they had previously with local varieties. The idea is that if a new line can really provide better yield under their existing conditions, no additional expenditure will have been involved and the farmer will have no difficulty in adopting the new variety/line when on-farm research projects in the village are discontinued.

Rather than working with a few selected farmers, whole villages are involved in these experiments. This avoids bias in picking out individuals, but the villages in which we work have been selected as consisting of resource-poor farmers with no irrigation. By 1987, a total of nine distinctly promising lines had been identified by farmers. This had been achieved within a much shorter period than normally necessary to produce only one official release and, whereas an official release might prove unacceptable when transferred to our resource-poor target group, those breeding lines which they have themselves listed and selected under their own physical and management conditions are by definition appropriate and acceptable to them.

1.3   Local knowledge for agroforestry and native plants

DIANNE ROCHELEAU, KAMOJI WACHIRA, LUIS MALARET, BERNARD MUCHIRI WANJOHI

Local knowledge and research processes in Africa

Agroforestry has become popular in development and environmental circles throughout the world. It is now often invoked as a new solution to rural development needs. But the scientific community and development agencies have not invented agroforestry; this is merely a new word used to describe age-old land-use practices familiar to millions of farmers and herders in many parts of the world.

Agroforestry is defined formally as a holistic approach to land use, based on the combination of trees and shrubs with crops, pastures or animals on the same land unit, either in sequence or at the same time (Lundgren, 1982). In reality, most farmers cannot easily separate this from the integration of woody plants into agricultural and pastoral landscapes. Whereas formal agroforestry science is based on the systematic placement of trees relative to crops and pastures, rural people are often more concerned about the fit of the whole agroforestry practice, and trees in general, into the larger landscape. In many cases, farmers have longer experience and knowledge of 'agroforestry' practices than scientists.

In Eastern and Southern Africa, the trend of most agricultural development and settlement programmes has been toward the oversimplification of production systems and 'homogenization' of landscape; in addition, such programmes have often accelerated the twin processes of resource degradation and selective impoverishment of women, the poor, and/or ethnic minorities. Even most existing agroforestry programmes suffer from an imbalance of technical and social expertise and from lack of accountability to their rural clients.

Only recently have we begun to undertake successful agroforestry (AF) programmes which depart significantly from these patterns. These emphasize the priorities, knowledge, innovative capacities and full participation of local people in research and development. Key attributes are adaptability to local conditions, adoptability by farmers, and genetic diversity. Where farming communities are already well established, these AF programmes face choices about the use of residual woodlands, the conservation of local knowledge about plant species and their environments, and the domestication of valuable wild species into cropland, pasture and other niches. Land use and landscape planning for ecological and economic diversity are also involved. The challenge is to encourage, support and supplement rural people's own innovations in ways which combine these elements.

This is made more difficult by two characteristics of AF. The first is decisions which are committing. The choice between varieties of maize from research stations is simply compared with choices about land clearing, land use, tree planting and the management of woodlands. Farmers can change maize varieties after one season, but decisions about woody plant resources and the soil are not so readily reversible (Wilson 1987). Decisions in the present may determine resource conditions for generations to come.

The second difficulty with AF innovation is the enormous range of species and AF practices. Scientists lack proven packages for the diverse environments and circumstances of rural people in the region (Rocheleau and Raintree 1986). One response by AF researchers has been to choose a few practices and a short list of species and test them under a variety of circumstances, yet there is shortage of time and resources for such a trial-and-error approach to AF research.

15

For both practical and ethical reasons, the rural poor should predominate in these complex processes of technology and land use change. Practically, formal experiments have limited scope because trees require a lot of space and a fairly long time to grow. Only a few formal experiments can therefore be carried out, with few repetitions in space and time. Nor can formal testing be undertaken on a scale to fit the numerous, distinct environments which are commonly found. Great care is consequently needed in deciding what species and what AF practices to submit to formal experiments (Raintree 1983; Huxley and Wood 1984; Torres 1984). Moreover, the complexity and scope of the changes involved are beyond the capacity of formal research programmes under controlled conditions.

Rural people have here a comparative advantage: they know and use whole systems in all their diversity and variability; as clients they know what will meet their needs and they are well placed to adapt and adjust AF components over time. From an ethical point of view, also, it is right that the poor rural majority should direct any process which will transform the rural landscape and the biological basis of their livelihoods.

AF research and development workers in the field must therefore carefully mix existing local practice with the science of designing and testing new practices, involving themselves as consultants and catalysts in a process of research, extension and evaluation essentially 'owned' by rural people (Rocheleau and Weber 1987).

Rapid Rural Appraisal (RRA) and ecological methods for community-based AF research

The methods for agroforestry in general and community-based AF research in particular must constitute a radical departure from traditional agronomy and even from many of the farming systems research methods that have become established in formal scientific circles. Whether in formal or informal research programmes the approach should often be more ecological than agronomic, as befits the focus on the place of trees, woodlands and savannas in the habitat of farmers and herders. Within ecology, both qualitative and quantitative sampling and monitoring techniques have been developed to study whole systems and the complex relationships between organisms and their environments (Odum, 1984; Conway, 1985). Moreover, the theory and the methodology are well suited to a sliding scale of analysis from tree-soil interactions to regional land use systems (Odum, 1984; Rocheleau, 1983; Hart, 1985; Conway, 1985), whereas agronomy is firmly rooted in the plot.

The development of AF and woodland management systems for rural landscapes can benefit particularly from the convergence of methods in two sub-fields of ecology – ethnobotany and agroecology. While ethnobotany draws its methods from human ecology and ethnographic traditions in anthropology (Posey, 1981; Nations 1982) and naturalist traditions in plant and animal ecology (Okafor, 1981), agroecology derives its research methods more from environmental management and systems ecology (Hart, 1981; Altieri, 1983; Conway, 1986). Ethnobotany and agroecology

16

provide tools for studying existing 'natural' ecosystems, traditional AF systems and recent innovations by rural people. Their methods present ample scope for incorporating indigenous technical knowledge, indigenous capacity for innovation and indigenous capacity for experimentation into the identification of species for domestication and the design and testing of new AF and woodland management systems.

Rapid Rural Appraisal (RRA) techniques can combine readily with ethnoecological methods. However, it is the style rather than the speed of RRA which is most critical. For example, researchers can nest ethnoecological data and sample collection methods within a series of informal interviews with rural community groups of 15 to 30 people, followed by 'chains' of household level and individual interviews, mapping of farms and collection areas and participation in gathering trips, processing and other activities. During subsequent stages of research the same kinds of information-gathering activities can be used for monitoring and evaluation of experiments, whether formal or informal. This can apply whether the experiments are on-station, on-farm or in-the-forest, over a wide range of 'user'-and-'researcher' partnerships with respect to experimental design and management.

The possibilities range from research-designed experiments on-station to rural people's own on-site experiments that are simply 'discovered' and documented by research institutions. Most programmes are based on a more direct collaboration between the two groups, which includes a variety of roles for land users and formal research institutions in experimental design and management (Feldstein, Poats and Rocheleau, 1987).

Most of the immediate work in community-based AF research will focus on ecological adaptations of RRA combined with experimental situations where the user is also a researcher. However, the exact choice of methods and how to combine and apply them is still largely a matter of taste, style and available resources. For most professional researchers, first attempts with such an approach will be somewhat of a personal experiment to derive a coherent methodology from an eclectic collection of methods to answer research questions framed in response to local circumstances.

The two cases which follow are not models, but examples of such experiments. The emphasis is on lessons learned, and implications for follow-up.

An example from Kenya: trials, errors and hindsight

In exploratory on-farm research conducted by ICRAF in Mbiuni Location, Machakos District, Kenya, we tried out a combination of the methods described above. An earlier project in the area was based on the Diagnosis and Design (D&D) method (Vonk, 1984; ICRAF, 1983 a, b, Raintree, 1983) and involved a farm-level survey, on-farm AF trials and monitoring of local farms. In previous cycles of diagnosis and trials, farmers had identified priority problems for AF research to address: poor soil fertility, inadequate soil moisture, dry season fodder shortage and lack of building material and fuelwood. The proposed responses to these concerns included

17

alley cropping with *Leucaena leucocephala* for mulch and fuelwood and rehabilitation of grazing lands through planting scattered multipurpose fodder trees in microcatchments. Ten farmers tried some combination of these in informal trials which were on-farm, researcher-designed and farmer managed (Rocheleau, 1985).

Later work at the community level and follow-up of the original ten farm trials provided a wealth of information and innovations based upon first, involvement of self-help groups in tree propagation and planting, second, participating farmers' reactions and proposed alternatives to the original technology trials and third, reaction of the group members to their own tree planting efforts on-farm and to the original ten trials (Rocheleau, 1985). The researchers joined self-help groups as participant-observers in weekly soil conservation sessions. The researchers proposed AF practices to supplement structures at gully and grazing land rehabilitation sites (Hoek, 1984; Rocheleau and Hoek, 1984), but participating farmers requested seedlings for on-farm planting rather than 'wasting' them on the conservation sites. At planting time project staff obtained seedlings from a government nursery and distributed 'sampler packages' of 13 exotic tree species to 120 active participating members of five collaborating self-help groups. Each participant had agreed to allow follow-up surveys and to observe and report on the performance of the trees. Other members of the community expressed interest in securing seedlings for the next planting season and within a few months six groups asked help to develop small nurseries and to grow their own seedlings (Rocheleau, 1985).

The new trials by group members were informal and exploratory and often incorporated either the function or the form of the alley cropping and grazing rehabilitation technologies, but rarely both. When group members were invited to visit and discuss some of the original on-farm trials as a group, they shared their very critical opinions about the 'package' in question, and also 'adopted' the process of AF development as a community enterprise.

By explaining to farmers the intentions and reasons for the trials researchers gave the participants a basis for assessing them constructively rather than simply accepting or rejecting them. During the course of the discussion participants raised several critical points about the trial technologies, which led others to pose alternative AF designs. For example, of the group representatives who visited the alley cropping site, one woman was struck by the attempt to improve soil fertility through the addition of plant biomass (mulching). Soon afterward she approached the local farmer-extensionist-researcher (and host of the alley-cropping trial) with her own practice of 'boma-mulching'. It consisted of applying large amounts of bulky plant biomass trimmed from living fences of *Euphorbia terucalli* to *bomas* (cattle pens), to be soaked with urine, trampled by cattle and baked into the underlying manure and soil. This produced instant compost.

As the boma mulching practice was discussed in group meetings, more people indicated some experience with the technique and many more showed an interest in trying it. Others reported having used *Terminalia*

18

brownii and *Combretum* spp. leaves from large dispersed trees, and had been doing so for years. Within the year most farmers in the vicinity had tried this at least once after trimming their *Euphorbia* hedges. The next logical step seemed to be refinement of the technique in order to increase the nutrient content and to increase bulk without fouling the cattle pens.

Another woman who was present at the same discussion at the farm trial site planted three species of fruit trees in lines at 4m intervals in her cropland and a checkerboard pattern of *Leucaena* in her vegetable garden for wood and mulch. Yet another member who visited and discussed trials on the same farm, decided to plant a wood and timber lot on a degraded cropland plot, as well as living fences and timber on her property boundaries and a mix of fodder and timber trees in a small pasture near her home. Three others present at the same group discussion of the trials followed up by planting trees in cropland for fodder or for small poles.

The participating farmers saw themselves as choosing, mixing and matching from a selection of possible AF practices with some demonstrated feasibility. They were not adopting a proven package. The group participants began to request seeds and seedlings of particular species. As they gained more experience with tree propagation and planting *per se*, more farmers also began to come forward with experiences or interest or knowledge relating to indigenous trees. They also developed a keen sense of the vulnerability of some exotic tree species to drought, browsing, trampling and termites, which further fuelled their interest in indigenous trees.

Many of the group members expressed an interest in learning to grow local species. In one case women's group members asked for more plastic tubes for seedlings and researchers asked if farmers could provide local tree-seeds in exchange. This set off an animated discussion, since many of the participants had assumed that project researchers only dealt with exotic trees.

During the course of this group discussion two elder women recounted having tried to grow *Acacia tortilis* and *Balenites aegyptiaca* and having failed, which was determined to be from lack of seed treatment. The whole group welcomed the subsequent discussions and demonstrations of seed treatment for both indigenous and exotic species. The same group later collected their own seed of *Acacia polycacantha* from a tree on the group leader's property, treated the seed by two methods, and raised approximately 200 seedlings for planting by group members during the next rains.

As more species became available in projects and nurseries, some farmers began to trade and barter with trees. Observation of this trading activity, as well as the subsequent use of the seedlings, revealed a wealth of information about *who* wanted *what kind* of trees, for *what purposes* and *where* they were willing to put them.

As the follow-up and additional distribution programmes proceeded over the next two years, farmers became increasingly aware of the importance of termite and drought resistance. At the same time more and more trees, both indigenous and exotic, were being planted by farmers in their gardens, protected croplands, fencerows and close to the home

19

compounds, as people developed awareness of the advantages of having trees in those areas.

There were many other examples of farmers' inventiveness, experimental successes and productive interactions with researchers through the use of these interactive research methods. The group level activities also resulted in a transfer of tree propagation and planting technology from the hands of a few skilled and relatively well-off men to most farmers in the community, the majority being women. Their very involvement in these activities changed the species, spaces and processes which emerged as part of the evolving research agenda.

Out of all the initiatives taken and questions posed, several potential research directions emerged. The farmers' priority interests included the use of plant biomass for soil fertility, the use of leaf mulch from dispersed trees outside the cropland and multi-storey systems for land use intensification. They also adopted the process of AF development and domestication of trees, incorporated timber species from earlier trials and sought solutions to their own specific tree-planting problems. This list of priorities is distinct from the formal (or conventional) scientific sequence of:

- species selection and genetic improvement of plant material;
- development of prototype technology;
- adaptation of prototype to sites; and
- widespread extension of a fixed package.

By contrast, this experience argues for introduction of many varieties or species and a few sample technologies with emphasis on principles and demonstration of some promising components – as effective approaches to help build sustainable R&D processes for resource-poor people.

Plant domestication: local knowledge and 'chain of interviews'

We were particularly interested in women's use of off-farm lands, which included the gathering of indigenous plants and the appearance of more and more 'incipient' home gardens. We also wanted to help develop alternatives that would serve women most dependent on products gathered off-farm. The project started with identification of species and spaces most important to women gatherers and investigating their interest in domesticating favoured wild species on farms or in managing woodland systems. We focused primarily on food and medicinal plants and secondarily on wood fuel and fodder plants.

We used several methods to describe the existing situation with respect to the role of wild indigenous plants in land use systems and document traditional practice and local knowledge, identifying recent innovations in plant management. The effort relied heavily on informal surveys of groups, household and individuals among both the community at large and acknowledged local experts (Pope, 1986; Rocheleau et al, 1986; Malaret & Ngoru, 1986). In particular, we developed the 'chain of interviews' method.

This started with interviews and group discussions built on prior contacts

20

from earlier farm trials and group activities in soil conservation and tree planting. In these meetings, the purpose of the research and range of topics and specialists were identified. This led on to household interviews and lengthy talks with local specialists. These encounters in turn often led into participant observation on gathering trips, visits to sites of tree-planting or plant domestication and longer-distance travel to special collecting locations (Rocheleau et al, 1986; Wanjohi, 1987; Wachira, 1987). Researchers also conducted opportunistic interviews when they happened upon people herding animals or gathering food, medicine, or fuelwood. The residence of researchers in the area also provided opportunities for farmer-initiated interviews and information exchanges.

The group discussions normally lasted about an hour, with 15 to 30 people present. The early meetings entailed listing of plants gathered and places used for particular products. In later sessions the group discussed reasons for practices and preferences, problems with plants and source areas and ideas for improving the situation; eventually the group tackled decisions about which plants to domesticate and where and in what combinations. Interviews often ended with questions for participants to consider, followed in a few days by another session which gave people time to think and to confer with family and friends (J Kyengo, personal communication; Vonk, 1986; Rocheleau, 1985).

The household and individual interviews varied in time and in format, depending on the disposition of the persons involved. Both formal and informal approaches were used. One informal in-depth survey was based on a chain of informants from 'average' to expert; another was a more formal randomized sample of 63 households (5 per cent of population), which asked farmers to answer specific questions about the environment, collection, use and preferences of wild plants, etc. (Mutiso, 1986; Wanjohi, 1987; Munyao, 1987; Wachira, 1986; Rocheleau et al, 1986). The formal survey took three times as long and reproduced the same main results as the group interviews and chain of interviews, with less detail and coherence.

The surveys on women's use of off-farm lands and gathered plants yielded a list of 65 indigenous species used for food and 99 used for medicinal purposes, among them woody species, wild leafy vegetables and wild roots (Rocheleau et al, 1986). Most of the fruit-bearing woody species were also major sources of wood or fodder, uses which had received more attention in previous surveys of the farming system. In the formal survey 90 per cent of the 5 per cent sample group reported using gathered leafy vegetables to some extent, 10 per cent said they use wild greens year round and 70 per cent reported that they or their children eat wild fruits daily (or whenever available). Most of the respondents also used herbal remedies made from indigenous plants.

In many cases people noted that wild plants play a particularly critical role at some times of the year. Some of the wild greens, such as *Commelina africana* (Kikowe), are particularly important for late planters (ie, poor people who 'borrow' or rent oxen) since these greens fill the gap between the onset of the rains and the first harvest of cowpea leaves from the cropland. Likewise, some vegetables (*Solanum nigrum* and *Amaranthus* spp.) and fruits are especially important during the dry season, with over 25 species of fruits used by the sample group during that time (Wanjohi, 1987; Mutiso, 1987; Wachira, 1987).

Of all the species listed, farmers identified four species of leafy vegetables, nine fruit-bearing species and seven medicinal plants as good candidates for domestication on-farm. The criteria cited for choice of candidates and the suggested planting niches and plant combinations also helped to define useful criteria for subsequent screening of exotics and for design of AF practices with both indigenous and exotic species.

Most women surveyed were interested in alternatives to the current situation of gathering products in degraded and sometimes distant collection areas. They were receptive to the domestication of indigenous trees (including wild fruits) and wild leafy vegetables in gardens, small tree plots near the home and in-between spaces such as boundaries, gullies and along drainage and soil conservation structures (Rocheleau et al, 1986). Most participants were also eager to try exotic species to supplement indigenous fruits and vegetables or to sell as cash crops to urban consumers. There was an especially keen interest in exotic *Amaranthus* species for home use as leaf spinach (Wanjohi, 1987). Athough they acknowledged the production potential of managed woodlands and grazing land, the people surveyed were overwhelmingly pessimistic about woodland management except for those who actually own sizeable chunks of land with bush and woodland vegetation (Wanjohi, 1987; Wachira, 1987).

Throughout this cycle of surveys and plant collection the tree planting extension continued, with substantial informal feedback to the research effort through this activity, as well as through participant observation in group work sessions. Moreover, the surveys sparked new interests which in turn fed back into extension and informal trials. Similarly effective group methods and processes have been woven into CARE agroforestry projects in other parts of Kenya (Vonk, 1986; Buck, 1987).

In Kathama several farmers (all women) who had participated in interviews requested assistance with the design and establishment of small home gardens for plant domestication. They also requested help in procuring seed and cuttings of indigenous trees, vegetables, and vines, as well as exotics. Each of the farmers planned her own garden (Wanjohi, 1987; Wachira, 1987), chose her site and cleared, tilled and fenced her plot prior to planting time.

The factors which influenced selection of indigenous wild species for the gardens included abundance, ease of access, and palatability (for both fruits and vegetables). Also important as selection criteria for vegetables were preparation requirements (ie, whether they need to be fermented in milk, fried in oil or boiled) and whether they are used alone or mixed into staple dishes as a relish.

Of the seven garden trials established during this season, five 'succeeded'. Success, as defined by the farmers meant that gardens produced enough green vegetables for home consumption or at least enough to reduce the need for gathered or purchased greens, or that gardens produced vegetables that were more palatable and easier to prepare than the usual mix of gathered greens.

Fruit trees were considered to be a tentative success if they established well without serious damage by pests, diseases or drought. Another measure of success was the degree of interest expressed by self-help groups and individual farmers in these gardens. Their priority interests included: development of vegetable gardens in homesteads and group sites; homestead fruit trees; mixed tree nurseries; multi-storey home gardens. They did not express these interests in terms of a fixed 'package' but mentioned ways to use the principles and components involved.

Problems cited (in different degree) were mainly browsing by livestock, insect pests and drought, which mirrored many of the difficulties experienced in earlier alley cropping and grazing land trials. However, given the size and location of gardens, farmers found these problems much easier to address. The home garden also presents a lower risk environment for experiments and allows farmers to observe the entire system close at hand. Yet, even in this limited and well defined space, farmers must deal with many related innovations (such as effective fencing, pest control, intensive soil management and research and training in plant propagation techniques) that are necessary to support new plants and practices.

Eventually several principles and components from these garden 'trials' will likely find their way into cropland, grazing land and in-between spaces in the larger landscape.

Lessons and follow-up questions

Several lessons were learned from the studies. The substantial interest in domestication of indigenous wild plants (including trees, shrubs and herbaceous plants) warrants vigorous research and extension to follow-up on those species identified for domestication or for management in place. The experience points to the practical value of:

- choice by farmers of indigenous and exotic species for AF systems, according to criteria identified by them;
- identification of source areas and screening of germplasm for farmer-selected species;
- testing propagation techniques for selected indigenous and exotic species for AF systems;
- conducting 'social' experiments with different tenure arrangements, and
- testing different technology designs for 'interlocking' land uses by multiple users at shared sites.

One methodological question is how the surveys should differ in timing, format and/or content if they were done again. One need is a broader base of ecological information and careful identification of topics for separate treatment and for systems research. An important gap in information is a summary of indigenous knowledge and environmental perception. A general ecological survey would help at the start to provide better understanding of communities' views and an empirical basis for analysing

23

ecological relationships between different land-use systems and arrangements of trees.

Systems level experiments by and for land users are also crucial. The home garden and small grazing land plots are the two niches most often chosen by farmers for trials of complex interventions. The particulars of research designs and types of trials will vary, but need to be planned carefully.

In all the suggested follow-up activities, the role of local participants will be critical. To be effective, continuing research with land users must rest on shared information and understanding; success of follow-up will depend on the strength of the partnership between the community and the field research team.[3]

1.4 Scientists' views of farmers' practices in India: barriers to effective interaction

ANIL K GUPTA

Researchers' attitudes[4]

In a recent study based on field work in semi-arid parts of Western India (Gupta, Patel and Shah, 1987), we asked a number of biological scientists to narrate any farmers' practices which had intrigued them. Our purpose was to understand whether the scientists, often blamed for ignoring farmers' innovations, were really unaware of them.

The sample included 61 scientists (24 from the All India Coordinated Research Project on Dryland Agriculture – AICRPDA, Hyderabad, 24 from Haryana Agricultural University, Hissar and 13 from the University's Dryland Research Station at Bawal). They were from different disciplines, ranging from plant breeding, genetics and agricultural engineering to agroeconomics and sociology. The main method of eliciting information was to interview the scientists with the help of a structured schedule of questions.

Several variables may influence the way in which a research community perceives the knowledge of peasant farmers, including the scientists' values, assumptions about the nature of scientific knowledge, dislike of simple technological alternatives, unjustified assumptions about the farmers' constraints and opportunities (Sanghi, 1987) and the urban 'tarmac' and related biases identified by Chambers (1983). In this study, the ecological background from which the rural-born researchers came was also a factor, as was professional training and disciplinary background.

In this sample, 24 of the 61 scientists did report unusual practices by farmers, and it was striking that these were mostly scientists from Hyderabad. Their observations can be classified as:

24

- sceptical;
- critical of the practices considered sub-optimal, or
- unscientific; and
- acknowledging that the practices are useful and innovative.

A limited number of examples are given in Table 1.1.

Table 1.1: Some typical responses of scientists regarding farmers' innovation and practice

---

Cryptic answers and sceptical comment

- 'Old farming practices are ... well tested. But a practice which was simple and good 50–100 years back may not be good in present circumstances. And a practice which is good today may not be good tomorrow. Every farmer's practice needs to change ... with time.'
- Farmers 'are not maintaining any record so when asked for previous practices they couldn't recall the actual performance, eg how much labour used in different operations'.
- 'I have not noted any interesting farming practice.'

Sub-optimal resource use or ignorant of alternative land use

- 'Use of less seed and fertilizer than the recommended quantity.'
- 'Moisture conservation practices are not adequate in rainfed areas.'

Apparently unscientific practices

- 'Some farmers do not till the fields during the fallow winter season because of the belief that soil will catch cold if ploughed then.'
- 'Sowing of seed of some crops mixed with fertilizers.'
- 'Farmers follow up-and-down cultivation methods without consideration of slope, so there is a risk of soil and water erosion.'

Acknowledged as innovative practices

- 'Growing of sarson (mustard) in criss-cross sowing in the gram crop.'
- 'Use of NAFE (a desi (traditional) plough) for deep sowing of gram by camel.' (Also other uses of traditional ploughs for sowing gram and mustard.)
- 'In some villages of Hissar and Sirsa districts farmers use a blade hoe for preparation of the seedbed. It is a very useful implement as it saves time, labour and at the same time conserves moisture.'

It should be noted that these scientists have rarely investigated the reasons for the practices they mentioned. Thus the science underlying rational practices and the myths behind not-so-scientific practices have not been understood. We want to state unambiguously that the mere

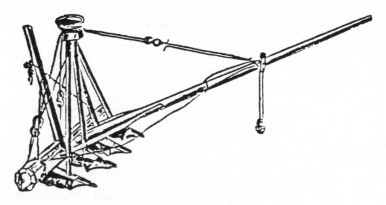

Figure 1.2: *Four-row seed drill as used in South India in the eighteenth century. This example was sent to England in 1795 or 1796 by Captain Thomas Halcott with the comment that, 'here is a remedy for the defect complained of in the English drill plough'.* The illustration was published in Communications to the Board of Agriculture, *London, 1797 (Dharampal, 1971: 211, 213)*

documentation of peasant practices is not enough. We have to identify the scientific basis of peasant practices and link it with their rationality. This view is linked to our plea that *science* should be transferred to the farmer and not just *technology*, so he knows the reasons for what is done and is better placed to improve his practice.

The importance of this can be illustrated by reference to the state of Indian agriculture in the eighteenth century as recorded by British travellers. One of them commented on the bullock-drawn seed-drills used in India (figure 1.2), noting that seeds were spaced at definite intervals along the rows. These implements surprised him because in England, the seed-drill was regarded as a recent invention, over-complicated and still unreliable. The Indian device had been in use for longer and was much more satisfactory (Dharampal, 1971). One reason why this technology did not advance in the East as rapidly as it subsequently progressed in Europe may well be the lack of a link with science – or at least the lack of interaction between farmers and those who wrote books about botany or agriculture. Indeed, the ground rules for classifying peasant knowledge and linking it with scientific method still need to be developed.

Extensionists' attitudes[5]

In a workshop of scientists and extension workers organized by Nurul Alam and his colleagues from the Bangladesh Agricultural Research Institute (BARI) and Rafique Ahmed of the Directorate of Agricultural Extension, Bangladesh, extension staff were asked to list those farming practices which they considered most intriguing. They were not to judge the efficacy of these practices, since it was acknowledged that they would all need to be tested before being considered worthwhile. A wide range of practices emerged, as indicated in Table 1.2.

Table 1.2:   Farmers' beliefs and practices reported but not tested by staff at the Department of Agricultural Extension, Tangail, Bangladesh

No	Staff member reporting	
1	Subash	Opium insertion in bottlegourd stem increases the number of fruits
2	Feroz	Non-bearing papaya bears fruit when injected with cholera vaccine
3	Feroz	If non-bearing bottlegourd vines are given a longitudinal incision they start bearing fruits
4	Awlad	Sowing of jute after the full moon in chaitra is considered optimal
5	Awlad	Powder of neem fruit used in paddy to control pest
6	Murshid	Urea is used for controlling stemborer in Boro paddy
7	Hoque	Broadcast ash over paddy to control insects
8	Sayed	Juice of Talakachi leaf mixed with water and sprinkled on leafy vegetables helps in the control of beetles
9	Alam	If jute is grown after wheat, a nodular substance in jute roots leads to mortality of the seedlings
10	Murshid	Laddering in wheat increases tillering at 20–25 days after sowing
11	Feroz	A longitudinal section cut after the dark phase of the moon of Bhadra or Shin in Jackfruit helps to encourage bearing of fruits
12	Alam	The banana plant is used for rat control in wheat (the rustling of leaves creates sounds which keep the rats away)
13	Awlad	If 'Shazna' cuttings are planted after the first shower in the chaitra, bearing starts within one year

We did not want the reporter of a practice to certify its reliability or the generality of its application, since this could have led many people to keep quiet. Under these conditions, even the District Agricultural Officer and other senior members of the workshop contributed, the philosophy being that if somebody could not cite any practice, then he had wasted his life!

We considered it impossible for anyone to work with farmers and not notice innovations or unusual practices at some time.

However, on the day following this exercise, we circulated the list of innovations with the names of the reporters, as in Table 1.2. There was strong scepticism when we announced that each innovation would be credited to the reporter. We noted embarrassed smiles on many faces when we actually did it the next day. Participants had probably thought that here was another snake charmer taking them up the garden path. This perhaps reflects the lack of credibility that we, the social scientists, have with grass-roots workers.

In a further exercise, Alam and his colleagues attempted to explore the beliefs and hypotheses which observed practices seemed to reflect. A list of these beliefs was incorporated into a questionnaire so that farmers could be asked with respect to each one whether they agreed strongly, agreed with qualifications, disagreed, or did not know. For example one of the beliefs listed in Table 1.2 is that the sowing of jute after a full moon in the month of chaitra is considered optimal, so a question was included about sowing time for jute relative to the lunar month.

The purpose of this work in Bangladesh was to get behind the myths associated with local knowledge and 'scout' for practical innovations. To this end, the scientists also had night meetings with farmers in the villages to discuss many of the innovations. They aimed to speak with older people separately from the younger generation in order to tap knowledge which the latter might despise.

Attitudes toward homestead gardens[6]

In another exercise which pained the biological scientists almost as much as the previous two, we tested their beliefs about the benefits of homestead gardens. The issue was that the horticultural department of BARI wanted to do a survey of homesteads in different parts of Bangladesh. They had drawn up a detailed questionnaire and consultants from an international centre doing research on vegetables had reinforced their view that only experts in the subject could decide what questions to ask.

However, when I was approached, I had to express helpfulness, but I also asked, how could one improve the questionnaire without under-standing the purpose of the survey, and the assumptions behind it? For biological scientists to be asked such things by a social scientist was disconcerting. Nevertheless, we had a meeting during which three state-ments were made very assertively by the scientists:

● households use the homestead space and other resources very inefficiently;
● they plant most of the trees, bushes and vegetables randomly, or just let the plants grow where they come up;
● they grow most trees for a single purpose, eg, fuel or fruit.

Once these assumptions had emerged, it seemed desirable to test them before moving further and a programme for doing this was drawn up consisting of the following steps.

28

First, a team of women scientists contacted a few poor women 'homestead managers' near the Institute.

Second, a map of all the fixtures on each homestead was prepared identifying each tree, vegetable and bush, spaces for tying animals, waste disposal sites, and so on. It was found that some scores of species of plants were grown (up to 70 species being noted in a later case study in Tangail District), with up to 40 being found at a typical household, including fruit trees, vegetables, herbs and shrubs.

Third, a discussion was held in which we asked why some species were found in greater numbers than others, analysing possible reasons in terms of the three coordinates of space, season and sector.

Fourth, multiple uses of different plants were noted in order of importance (eg, fruit, fodder and fuel from one tree).

Then a meeting was held in which those who spoke about the haphazard and random nature of the homestead biomass were asked to explain and interpret the homestead map. It was recognized, after long discussion, that the homestead planting was so complex that no firm conclusion could be drawn, with the available information. But there did seem to exist some order in what had been assumed to be disorder.

It was then decided to pursue a more detailed and more wide-ranging study with the help of some 28 women scientists from other divisions of the institute, most of whom did not normally work together. It was assumed that most decisions about homestead gardens were taken by the women in the homesteads and thus women scientists would have to help if adequate information was to be obtained, but it turned out later that the decision to leave self-sown tree seedlings intact, or alternatively to uproot them, was generally made by men, while vegetables and medicinal plants were tended by women.

The women scientists went out to different parts of the country to develop case studies and a large number of innovative practices emerged. Results were discussed with male field scientists, but it was suggested that gender-specific issues needed to be examined from a different point of view. The role of women in the homestead needed to be understood in terms of their own specialist knowledge and not just by regarding them as exploited workers who contribute to post-harvest chores. I consider the efforts by researchers on gender issues to spend major energy on finding out time-task allocation misplaced. It is the technical and institutional knowledge unique to women managers of homestead farms which must be built upon. Later there was considerable consternation because the results of this survey were analysed by male scientists without involving the women scientists who had done the fieldwork. What was still more frustrating was the decision of the concerned male scientists not to include these women researchers in designing a follow-up homestead management study.

The researchers' previous assumptions about inefficient land-use were then examined in relation to apparently unused space in homesteads which a horticulturalist thought could be used for growing vegetables, provided that they were watered during the winter. However, water was a scarce resource which was needed for washing cattle (who may get diseases otherwise), for personal hygiene, cooking and drinking as well as for irrigation of existing vegetables and trees. So not using water for planting extra vegetables was after all not a bad decision and certainly not a proof of irrationality.

Issues arising[7]

This exercise revealed barriers to interaction and understanding between senior scientists and extensionists, between female and male scientists and between biologists and social scientists, as well as between farmers and scientists. Aggravating the situation as far as the horticulturalists and biologists were concerned was pressure from donor agencies to start work without waiting for the results of the survey. This is mentioned to highlight the fact that in many developing countries, one should not assume that lack of innovative research is caused by lack of focus or proper methods.

A further implication of research with marginal farmers or women gardeners is the marginalization of the researcher himself or herself. We urge concerned scholars to note that since the majority of scientists are never likely to want to work with poor farmers, or become accountable to them, the dynamics of minority action need to be well understood. The minority who do work on these themes needs to be sustained in the short-run if better and more liberating forms of research are to emerge in the long run. If we ignore these issues, we will be in danger of creating a new myth of 'harijan scientists', who are close to God because they work with the poor, but who can be ignored for the time being because they are not close to God's creations, the men with power.

Finally, and most importantly, we must confront the ideas of the natural scientists who object to farmers' involvement in research but admit that the farmers do make some valid points. For example, it is partly justified to say that farmers are sometimes poor because they have not been able to innovate fast enough to keep pace with changing circumstances and the accumulation of new knowledge. It is also true that farmers who cannot read or write or keep records are handicapped when it comes to comparing current crop yields and labour requirements with what they achieved in the past.

However, the question which is most serious and which we hear most often from the senior agricultural research leaders is, have we not delivered the goods so far using our own methods? The implication is that more of the same will do. This misconception led many research planners to apply Indian wheat and upland irrigated rice experience in Africa without much success. Within India, the oilseed mission makes the same assumption. We are not questioning that some of the germ plasm which has been found good for high input environments may also be useful under stress environments, but we would argue that because the survival options of poor households in the latter area are so circumscribed, the research approach and agenda must be different. The success of modern wheat and rice varieties has resulted in some conceptual blinkers. In the different context of stress environments, where nature is more hostile and the demand for technological change is feebly articulated by poor farmers, work with the farmers becomes especially important.

A point which must not be missed in context of matching farmers' concerns with that of the scientists relates to the ability of farmers to *demand* what they *need*. Too much emphasis on responding to only the articulated demands of farmers may reduce the zone of responsibility of the scientists. It has to be conceded that farmers may not *demand* what they do not know or imagine can be supplied by scientists.

The limits of what scientists can do to help people in high risk environments need not be defined. It is not what people can demand but rather by what they *need*. Defining the needs of 'others' as well as one's own requires making value and moral judgements explicit – perhaps we have not done enough towards this.

1.5   Farmers' knowledge, innovations, and relation to science

IDS WORKSHOP[8]

Classification systems and their functions

There are far more examples of the innovations and local knowledge of Third World farmers than can be described in this book. The numbers of reports and articles on this subject reflect growing awareness of the value and potentials of learning from such knowledge (eg, Carlier, 1987; Farrington and Martin, 1987; IDS, 1983). Many of the scientists' comments about local knowledge concern rural people's classification systems for plants and soils. Local people use many categories in different parts of the world to describe types of land, landscape, crops, wild plant species and other natural resources. The categories and names used by them usually differ from those used by scientists. In addition, the criteria of classification are usually functional, that is, related to use, unlike the standardized categorization criteria derived from physical sciences.

In the semi-arid East Pokot region of Kenya, range management by pastoralists depends on how they draw distinctions between three kinds of grazing land – lowland areas used in the wet season, hill areas with perennial grasses and hills more specifically distinguished as reserve grazing for use in hard times (Barrow, 1987). Definable boundaries between these lands can be drawn on a map and related to ecological conditions. However, the point of this classification is that its basic criteria are *functional*, ie, related to use and so are different in kind from the ecological criteria used by scientists.

The highly discriminating land classifications used by many Zambian farmers are discussed by Stuart Kean (1987) and Richard Edwards (1987a).

The story is told of a European survey team in Zambia who made no contact with farmers and drew land classification maps which entirely omitted riverine strips and dambos (areas of retained moisture), two forms of micro-environment which, though small in area, are crucial for local farming systems.

Edwards (1987a) describes how farmers in one part of Botswana classify soils. Three main types, known as *mothlaba*, *mokata* and *seloko*, are distinguished on lands where sorghum is grown and Edwards comments on, 'a high degree of correlation between farmer classification and laboratory analysis . . . especially in the case of *mothlaba* and loamy sand' (Table 1.3). However, while laboratory classification is based on the relative proportions of sand, silt and clay in soil, farmers' classifications, as previously noted, tend to be functional. The fact that the farmers' *seloko* category seems to cover two kinds of loam soil identified in the laboratory is probably not the result of crude observation by the farmers, but arises because they are using different criteria, perhaps related to topographical conditions which affect ease of cultivation, or moisture conditions which influence crop growth.

Table 1.3 Classification of soil from sorghum sample plots in the Pelotshetla area of south-east Botswana by farmers and by laboratory analysis

Farmers' classification	Laboratory classification			Totals (%)
	Loamy sand (%)	Sandy loam (%)	Sandy clay loam (%)	
Mothlaba	18 (90)	2 (10)	0 (0)	20 (100)
Mokata	15 (19)	46 (58)	18 (23)	79 (100)
Seloko	0 (0)	8 (47)	9 (53)	17 (100)
Totals	33 (29)	56 (48)	27 (23)	116 (100)

Source: Flint, 1986; IFPP Farm Management Survey of Phase II; figures represent numbers of plots sampled.

Complexity of knowledge: the example of plant domestication in Kenya

Classification of plants by rural people may have a functional basis as well as their classification of soils. According to Calestous Juma (1987b), people in western Kenya group plants according to their uses and one functional name may then fit several species. Juma goes on to describe recent and

32

continuing efforts by 'resource-poor' farmers in Bungoma District to domesticate plants which had previously been collected from the forest. Several wild vegetables have already been brought into cultivation and it is likely that all the wild vegetables in regular use have now been domesticated. However, new domestications of fruit species are still being made, one example bearing what looks like grapefruit, but with the smell of bananas and the taste of passion-fruit. This was still, in July 1987, awaiting identification by the botanists.

In order to domesticate plants, people make trips into the mountains, usually on the Uganda side of the border to evade Kenya's strict forest laws and bring back plants and seeds to try on their own land. They first grow the plants in similar conditions to those obtaining where they found them. Thus a plant growing in moist ground near a stream would be planted in a similar position, but after such a plant has produced seed, farmers would plant them in conditions closer to those normally found on their land. The plants that survive and produce seed then have their offspring grown under normal farm conditions. The survivors from the second generation have thus been selected as a variety appropriate to ordinary farm or homestead environments. Sometimes, of course, the process fails, or more selections and experiments are necessary before domesticated varieties are successfully produced.

Juma (1987b) argues that to understand the evolutionary *innovative process* through which farmers introduce new genetic material into the economic system, it is not appropriate to use a reductionist methodology under which genetic resources are treated as something quite separate from their socio-ecological context. Thus although this is innovation which depends on experiment and on similar kinds of selection as used in 'scientific' plant breeding, this cannot be regarded simply as a case of farmers doing basic science. Nor is it sufficient to regard local knowledge of botany and ecology as 'technical' knowledge only, which implies that it is based on a western epistemology. Instead, the local knowledge about plants and seeds in this situation can be better understood if we recognize that it is based on a distinctive epistemology which is unique to these people's culture. Their knowledge and practices have co-evolved over time as adaptations to particular environmental, social, economic, and political circumstances and pressures. This epistemology is also associated with religious beliefs about the plants and their uses; in this sense, their religion is partly a way of treasuring valued resources. The complex belief system thus reflects the people's relationship to the ecology in this particular 'niche' in Kenya.

Several factors have combined to encourage the domestication of wild plants: the diversity of plant forms in the local ecology; social habits concerning resource utilization; influences for the introduction of cash crops; the presence of natural forest and its accessibility on the Ugandan side of the border. It should also be noted that the farmers responsible were once reliant on wild fruit and vegetables to complete their dietary needs, but with the extension of cash crops and the clearance of large areas of forest in the area, it is no longer so easy to gather fruit, roots, leafy vegetables and plants such as the aerial yam. Many families, too, are having to survive on smaller plots of land and are thus growing more yams and potatoes, which provide more food from a fixed area of land. Trees which can make productive use of odd corners of homestead plots and which provide fruit, edible leaves, fodder and perhaps firewood are also attractive. Domestication of wild plants is thus a response to complex pressures and opportunities by people whose access to the resources of commercial agriculture is restricted.

Juma (1987) also notes that the farmers in this area are, by nature, 'experimenters', in that they continually try out and adjust their practices and uses of plants in response to changing conditions. That is, 'a farmer is a person who experiments constantly because he is constantly moving into the unknown'.

Cropping patterns and innovations in Bangladesh

Much comment on innovations made by farmers centres on two kinds of change they may make in cropping patterns. First, they may discover advantages in new combinations of crops for mixed cropping, such as quick-maturing dryland rice grown with pigeon peas in India (Maurya and Bottrall, 1987), or passion fruit grown to shade coffee in Central America (Bunch, 1985).

Second, a kind of experiment which is frequently made with regard to cropping patterns is simply to vary the sequence of crops grown on one piece of land from season to season. Good examples of this are described by Hossain et al (1987), and were observed prior to and during on-farm research in Mymensingh district of Bangladesh. Here it is possible to grow no less than three rice crops each year, the main summer monsoon crop being known as *Aman*, the winter crop as *Boro* and the early monsoon or spring crop as *Aus*. Research findings suggest that in rain-fed areas, the greatest agronomic and economic advantage is obtained from a sequence of three crops, including wheat as a winter crop. However, on irrigated land, the recommendation is to grow only two crops of rice and to leave the land fallow when the *Aus* crop might otherwise have been planted. Having identified the advantages of growing two crops on irrigated land after some considerable research, Hossain et al (1987) comment:

> This practice was initiated by some innovative farmers even before there were recommendations based on research. More than fifteen years before, one farmer in Comilla district had observed that growing two crops in a year would be better than three crops even under irrigation (Bari, 1974).

Hossain and his co-workers then give a table – quoted again by Gupta (1987a) – to show that farmers are constantly experimenting with new cropping patterns. The same point is made here by Table 1.4 which shows that the area of land allocated to local (L) and modern (M) paddy varieties in the three rice-growing seasons varies greatly from year to year in the light of farmers' experience of yields, quality and resistance to pests and as

34

Table 1.4:   Percentage of area occupied by different crops in Kanhar irrigated site, Mymensingh District, Bangladesh.

	79–80	80–81	81–82	82–83	83–84	84–85	85–86
T. Aus (L)	44	–	–	–	4	–	–
T. Aus (M)	41	17	–	10	39	–	–
B. Aus	11	–	–	–	–	–	–
T. Aman (L)	51	44	28	17	30	–	6
T. Aman (M)	49	17	39	38	43	72	59
T. Aman (Pajam)	–	22	33	35	17	26	36
Boro (L)	61	44	72	33	7	–	2
Boro (M)	3	39	28	42	16	56	21
Boro (Pajam)	–	–	–	–	28	44	78

Abbreviations: The Aus and Aman rice crops may be either transplanted (T) or broadcast (B); the Boro crop is always transplanted.

Rice varieties are classified as local (L), modern (M) or Pajam, an improved variety distinguished by name.

(Percentages have been rounded to whole numbers from the two decimal places in the original).

Source: Hossain et al (1987).

they adjust to variable weather conditions during each year. Despite all this experimentation, however, the trend toward leaving land fallow instead of planting an *Aus* crop is clear. Since *Aman* is the main crop, the total area allocated to this amounts to almost 100 per cent every year, but local varieties are clearly being used less each year. However, the Pajam rice variety, recommended by researchers for rain-fed areas in the *Aman* season, was adopted in irrigated land in the *Boro* season without any such advice and has become popular, partly because its appearance and market price are similar to local varieties and its requirements regarding fertilizer are low.

Hossain and his colleagues conclude by saying that after seven years of pattern monitoring 'it is distinctly clear that very few patterns are stable and the farmers are continuously changing their cropping patterns'. This is rather disturbing when it takes scientists six to eight years to devise and test a new cropping pattern. During that time, 'the farmers will move far ahead of the researchers'. Thus scientists cannot easily 'solve' problems for the farmers. Instead they can make contributions to a continuing debate about solutions, in this case by designing several alternative technologies with which farmers can experiment and from which they can choose.

This latter approach is similar to Maurya's (section 1.2), but conflicts with the conventional wisdom in crop breeding and in most other technology

35

development, which is that the researcher's job is to produce an optimized solution to any problem tackled. The conventional wisdom, though, is based on the assumption that conditions are stable, so that once an optimum technology is introduced, it will remain the optimum. In reality, market, environmental and social conditions are in constant flux. Moreover, the resource-poor farmer is more vulnerable to such change than the wealthy farmer, who usually can more easily protect him/herself against climatic or market uncertainties. Thus while the wealthy farmer can often benefit from optimized technological packages, the resource-poor farmer must rely much more on his/her ability to experiment and adapt to changing conditions.

Nevertheless, this emphasis on farmers' capacities should not be taken to minimize the role of scientific research. Rather, it suggests a view of innovation as the result of a process in which farmers, researchers and others all take part. Hossain and his co-workers have made important contributions to the innovations they describe. Even though they limited themselves to offering farmers a range of alternatives, their figures suggest that crop yields in the area where they worked increased more substantially than would otherwise have been possible, especially on non-irrigated land.

Facts or superstitions: recognizing values and limitations[8]

One important question is how to deal with indigenous knowledge which is apparently tinged with myth and superstition. Although scientists may respect farmers' inventions such as a modified hoe which saves labour (Table 1.1), they are likely to ridicule the idea that the moon influences plant growth or that injections can encourage a plant to fruit (Table 1.2), because these make no obvious sense to them. Indeed, such ideas provide ammunition to scientists who are tempted to dismiss local knowledge as worthless.

Of course, it may be that conventional science has overlooked real processes whereby plants are affected by cuts, pricks or even injections in stems and by the gravitational pull of the moon. Folklore about the influence of the moon is reported from many parts of the world. In Central America, some farmers believe that when trees are being grafted, the grafts 'take' better and more quickly if they are done when the moon is at a particular stage. Similarly, many coffee farmers use the lunar cycle to determine appropriate timing for pruning their plants. Botanists know that the force of gravity partly explains why a seed in the soil pushes its shoot *upwards* and root *downwards* and also affects the flow of tree sap, so the lunar gravitational field could influence plant growth.

It may be a mistake, however, to think that every item of local knowledge about the environment conceals grains of scientific truth. Likewise, 'romanticizing' indigenous belief systems can be inappropriate since some mythical credence can lead to irrational behaviour. Yet, formal science insulates itself from large areas of life by disregarding everything which does not fit its categories of thought and much in human experience does *not* fit these categories, even though it has meaning in other respects.

Because the lunar cycle is associated in many people's minds with the menstrual cycle in human physiology, the moon can become a symbol of the way times favourable to fertility and growth alternate with times which are less favourable. This sort of symbolism can have imaginative appeal, regardless of any physical phenomena affected by monthly cycles. A person who undertakes a high-risk operation such as grafting trees or sowing seed in a region with very variable climate may derive reassurance from good symbolism associated with the occasion as well as from the practical precautions he takes. In countries where lunar calendars are still in use (such as Bangladesh), practical information about planting dates is expressed in terms of lunar months in any case. Thus references to the moon may indicate a very wide complex of ideas, mythical and symbolic as well as practical and 'scientifically' logical.

The limitations of indigenous technical knowledge (ITK) nevertheless need to be acknowledged. As noted by Swift (1979) and Biggs and Clay (1980) some of these are:

- indigenous technical knowledge and innovative capacity is unevenly distributed within and across communities;
- the ability of individuals to generate, implement, and transfer ITK varies greatly;
- social groups and economic stratification affect the type and extent of ITK in rural societies (eg, richer individuals are likely to innovate more in aggregate, but poorer people may be forced to innovate by their poverty (Swift, 1979));
- the transfer and use of information is sometimes constrained and error-prone where it has to be passed on orally and held in the heads of practitioners;
- the scope for improvements from 'pure' indigenous technical knowledge is limited to what can be done with the local pool of techniques, materials and genetic resources, plus whatever is introduced casually;
- many genetic possibilities are not explored within the informal system, such as the crossing of self-pollinating crops where specific plant breeding techniques are required;
- ITK may break down when people are faced with an environmental crisis or external interventions (Farrington and Martin, 1987).

Recognition of these limitations, as well as its advantages and values helps in taking a balanced view of ITK and of local people's innovations and of their potential.

Scientized packages or cultural integrity?

Interacting with farm people and building on their knowledge may require us to come to terms with the mythical/symbolic component of this knowledge, and in the next section, Paul Richards suggests a way of understanding some of the things which at first sight seem 'pseudo-scientific mumbo-jumbo'. But when researchers talk about 'legitimizing' indigenous 'technical' knowledge and encouraging farmers to use it, something more than understanding is required. Researchers also need to be clear in their own minds about whether they aim to legitimize local knowledge solely in the eyes of the scientific community, by picking out the 'tit-bits' of practical information, or whether they are trying to strengthen and maintain its cultural integrity. Juma (1987b) has argued that indigenous knowledge could be 'delegitimized' in the eyes of local people, or reduced to trivia, if isolated from its cultural context and forced into the framework of western epistemology. Thrupp (1987) similarly argues that simply to collect the technically useful items of local knowledge in 'scientized packages' will tend to devalue it.

Legitimizing local knowledge may be important in maintaining a people's sense of values and in opposing cultural threats from outside, but to achieve that necessary recognition by discarding aspects of knowledge which refer, through symbolism, to social values, is self-defeating and contradictory. For example, in parts of Kenya, pastoralists have their own system of range management based on extensive indigenous ecological and *social* knowledge. If the pastoralists are to retain their identity and lifestyle, they must make their range management knowledge seem rational and legitimate to the government, but it is difficult to do this without sacrificing the social and cultural content of the knowledge which is a large part of what makes it effective.

The problem is seen in its starkest terms in the context of indigenous medical practice. Physical illness always has social and psychological implications and the symbolism and ritual associated with much traditional medicine in Africa provides a means of coping with them. Traditional medicine is now achieving some recognition, partly by adopting professional organizations and partly with the support of authorities unable to reach all their people with conventional medical services. But according to Last and Chavunduka (1988:267):

> there is an inherent danger that traditional medical knowledge will be defined simply in terms of its technical herbal expertise, that this experience will in turn be recognized only for its empirical pharmacognosy, without reference to the symbolic and ritual matrix within which it is used – still less the social matrix in which those rituals and symbols have meaning.

The risks encountered in farming are neither so personal as in illness, nor usually so threatening, so the ritual content of agricultural knowledge and technique is usually less than in medicine. Even so, there is often a ritual content for reasons which the next section attempts to explain.

1.6 Agriculture as a performance

PAUL RICHARDS

Is R&D directed at the wrong target?

In the rice-growing zone of West Africa, much agricultural research effort since the 1930s has gone into varietal selection. Release and spread of improved varieties has been a key component in a number of subsequent 'green revolution'-type initiatives. Improved dryland rice varieties outyield local varieties by about 10 to 30 per cent in typical on-farm conditions.

The major constraint determining success or failure in the Mende village in Sierra Leone where I worked in 1982–3 was timely access to labour – especially access to cooperative labour groups during the rice-planting season. To secure a labour group at the right moment it is necessary both to command a range of social skills (to know how to 'beg' the convenors) and to be in a position to offer the group the right food and other perquisites.

Labour groups will down tools if the food is not up to standard. They must be offered rice. There must be fish or meat and sufficient salt in the stew. Alcohol, cigarettes and cola are additional inducements. The business of putting together an agricultural work party is not unlike the business of organizing a dance, the other kind of party which enlivens Mende village life. The parallel is especially close where labour groups work to musical accompaniment.

Agricultural researchers spend much time measuring rice yields, but there are few measurements relating to the significance of music in agricultural production. What is the impact of drumming on agricultural labour? In one case where I undertook measurements of the same group working on the same day with and without music, 20 per cent more work was done to drumming than without it. I find this figure intriguing. It relates to what I would term a performance factor and is but one among many instances in peasant farming in Africa where the difference between getting a performance factor right and wrong is of the same order of magnitude as the productivity increment to be had from adopting research recommendations.

By and large, agricultural research has so far ignored performance as an area for systematic enquiry. This is not for want of material. Much of social theory is a theory of performance. The ethnographic literature contains many relevant examples, not least concerning the connection between music and work, or brewing and the organization of work parties. The significance of this material, however, seems to have eluded agriculturalists working on small-farmer farming systems.

The meaning of 'performance' in this context can be illustrated by an example which also shows how distant normal agricultural research is from performance thinking. The example comes from a discussion by Michael Watts (1983) concerning the way Hausa farmers in a village in Katsina, northern Nigeria, compensate for the effects of poor rainfall. What he observed was that the farmers make a series of rolling adjustments to drought. If the rains are late or stop unexpectedly, the first planting of sorghum may fail. The existing farm is replanted as many times as is necessary or until the farmer no longer has any seed left. At each replanting a different seed mix may be tried, better to fit available resources to changing circumstances. As the need arises and resources permit the farmer may then hedge or criss-cross the main plot with various back-up and insurance crops.

Farming systems researchers might imagine themselves to be on familiar ground at this point. They would tend (so Watts argues) to treat each of these resulting cropping patterns as a pre-determined design, as if in effect each farmer had said, 'this year to minimize the risk of drought I will plant so much sorghum, so much millet, so much cassava', etc.

This is to misunderstand almost entirely what has happened. The crop mix – the layout of different crops in the field – is not a design but a result, a completed performance. What transpired in that performance and why can only be interpreted by reconstructing the sequence of events in time. Each mixture is an historical record of what happened to a specific farmer on a specific piece of land in a specific year, not an attempt to implement a general theory of inter-species ecological complementarity (as plant ecologists might suppose).

Researchers, then, are looking at the wrong problem. They are looking for the combinatorial logic in intercropping where what matters to the Hausa farmer is sequential adjustment to unpredictable conditions. It is important therefore not to confuse spatial with temporal logic – not to conflate plan and performance.

But conventional agricultural research is not good at coping with performance issues for basic methodological reasons. Trials and experiments are 'out of time'. This is the basis for replication and comparison. By contrast the issues at stake in performance only become apparent when the performance is for real.

Thinking about performance

If conventional agricultural R&D has so far failed to take on performance issues, where might we look for models and inspiration? Musical performance is not a bad starting point, not only because music is integral to agricultural performance in many societies, but because it provides some useful questions about the link between analyst and performer.

A useful parallel can be drawn between musical analysts (critics and scholars) in 'western' concert music and agricultural scientists. Both are high status intellectuals concerned to understand how their subject matter works. The analogy breaks down (in a useful and thought-provoking way) when we factor in the performer. Concert artists are at least the equal of musical analysts in power and social standing. The connection between 'research' and 'performance' is open to negotiation between equals: some performers find analysis helpful and interesting, others are openly sceptical about what musicology contributes to their success as performers.

Agricultural research for resource-poor farmers is different. Here the performers are all of low status and little influence. They too may be sceptical of whether research helps, but they have little scope for voicing this scepticism. In this case, analysts are powerful individuals whose confidence that performers would perform better if they hearkened to analytical advice brooks no argument.

Chambers (1983) has addressed this asymmetry between analysts and performers in tropical agriculture and has suggested dealing with it by a series of conscious inversions and role-reversals – trying to get researchers to assume the farmer's standpoint. One way to do this might be to impose 'real life' constraints on the running of experiments and trials. This, I take it, is one of the factors in recent enthusiasm for on-farm trials and with-farmer research programmes. Trying to run a farm with the resources available to the typical peasant farmer is certainly a salutary experience. I would argue, however, that such initiatives will remain unrealistic from the performance point of view because they are powerless to grasp the way in which farming operations are embedded in a social context and therefore miss the contingencies generated by that context (reasons of the 'last week we had to sell the cow to pay for granny's funeral' kind).

This is something with which musical performers are familiar. They study, analyse, practice not to make mistakes. They plan ahead how to phrase a melody, coordinate entrances, pace the various sections of a piece, but much of this planning may go awry on the night. Faced with the realities of an audience it suddenly seems different. A good musician needs other skills, therefore – how to overcome nerves, how not to panic, how to recover from mistakes. No one, however talented, plays perfectly all the time. The capacity to keep going and avoid complete breakdown is always an important musical skill, however hard to define or teach.

It may be of interest, therefore, to agriculturalists to pay systematic attention to the coping skills of concert performers. An initial survey suggests the range of strategies is unusually wide. Some are based on common sense and experience. Others depend on medication or advice from psychologists. Then there are those based on 'indigenous' theories developed by performers themselves. Much in the last category will appear to outsiders to be pseudo-scientific mumbo-jumbo. But to the performer grappling with nerves and stage fright, scientific respectability is of little significance. It only matters that it works.

This helps, I think, put much 'indigenous technical knowledge' in the agricultural field into a new and useful context. Much of it should be judged and valued not by the standards of scientific analysis, but as self-help therapy through which farmers put their mistakes and disasters behind them without the performance grinding to a halt. But to treat ITK as a patch and mend philosophy in this way is not to devalue it. The problem is that science (infatuated with endless vistas of new research funding?) totally underestimates the capacity to keep going under difficulties. In the appalling environmental and economic conditions faced by many poor farmers in the tropics even to reproduce the status quo is often a brilliantly innovative achievement.

Perhaps the gap between farmers and researchers could be closed if those on the formal side of the fence reflected upon one lesson in particular from the musical field. Technical perfection is no guarantee that an audience will be moved. Conversely, technically imperfect performances are sometimes great performances. The composer Gustav Holst (reflecting upon musical performances by amateurs) used to say that 'if a thing is worth doing at all it is worth doing badly'. This comes close to the essence of what it is about performance that so frequently eludes 'normal science'.

Implications for research methods

How might agriculturalists begin to understand agriculture as social action and determine new (though inevitably more modest) targets for assistance to agricultural activities inextricably bound up in larger social processes?

One answer is that so-called ethnographic methods will assume much greater prominence in agricultural research than hitherto. Ethnographic methods (notably participant observation) allow some access to and understanding of performance issues in agriculture. The approach is not new. It was notably pioneered by de Schlippe (1956), an agronomist who retrained as an anthropologist and wrote what is still one of the best books on performance in African agriculture. One of his great achievements was to show that aspects of life totally alien to agriculture in a scientist's eyes are eminently explicable when seen in performance terms. One example is the relevance of witchcraft beliefs in the process of screwing up the performer's nerves to 'concert pitch' (or alternatively, undermining the confidence of rivals, perhaps deterring thieves from raiding isolated farm encampments during lengthy dry-season absences on hunting expeditions).

The attention paid to participants' own theories of performance is a central feature of the ethnography of performance. Again, some of the best material concerns music, notably in Ruth Stone's (1982) book on the organization of the musical event among the Kpelle of Liberia. She pays particular attention to the way in which sponsors of musical events, musicians and audiences, negotiate a performance and then how they understand the business of performing well. This introduces the reader to a range of performance skills, as understood by the Kpelle – timing, turn-taking, how to begin and end, how to cue, entrances and exits, how to cope with mistakes and broader notions of harmony, togetherness and the social and spiritual auspices under which music takes place.

Stone's study is especially interesting when read alongside the work of Bellman (1984) on the social uses of secrecy in Kpelle society. Bellman, working within the ethnomethodological tradition, is concerned with the way the Kpelle use ideas about ritual secrecy to segregate and demarcate distinct discourses. The ability to speak in Kpelle is far from being simply a question of possessing relevant knowledge. 'Speaking' is having a licence to perform. Such licences are gained through membership of appropriate closed associations ('secret societies').

This is a useful and immediate corrective to any naive view of the possibilities for interaction between farmers and agricultural scientists, or to simple belief in the capacity of such dialogue to achieve generally beneficial results. Researchers would first have to examine the auspices under which any participatory debate took place and how those auspices were interpreted both by participants and bystanders. Since it is not obvious without careful prior empirical investigation that Kpelle notions on these points would in any way coincide with those of agricultural researchers, the possibilities for cultural mis-communication must be enormous.

Thus accounts of agricultural performance informed by critical insights of the kind deployed by Stone and Bellman are badly needed in agricultural research. As my material at the outset suggests one place to start would be the process of labour negotiation. Another is how 'household farming units' are put together. 'Farm households' are not fixed in social structure. To a large extent they are the result of specific social negotiations (eg, marriage transactions). In some cases, they are negotiated and renegotiated on an annual basis. This brings into question the tendency among agricultural economists and farming systems researchers to treat the 'farm household' as a unit.

Another obvious area for further work is performance under duress. Coping skills in agriculture are often especially difficult to pin down systematically and describe, but there have been good beginnings in the work of Michael Watts on coping with drought and Barbara Harrell-Bond (1986) on refugee resettlement. This last study is especially important for demonstrating the extent to which refugee survival is skilled social achievement. By describing the contrast in fortunes of self-settled refugees and those in camps run by agencies, Harrell-Bond demonstrates the need above all to sustain those senses of vision and purpose through which social groups retain their capacity to act in a creative and cohesive manner.

1.7   Interactions for local innovation

IDS WORKSHOP[8]

Scientists and farmers

The theme of interactions runs through this book. Effective research and development by and with resource-poor farmers requires understanding and interactions of many types and at many stages. This includes social relationships, exchanges of ideas and information, linkages between people, and institutional dimensions (to be considered in Part 4). Here we consider interactions between researchers and farmers, between extension workers and farmers, between women and men and between outside science and technology and local capacities.

The importance of the farmer-scientist link has been illustrated by the examples already described; and methods used to improve this relationship for effective R&D programmes to benefit the rural poor will be discussed throughout the book. Farmers' groups and workshops are one useful way to help elicit farmers' ideas, to improve communication, and to foster local initiatives, and will be discussed in Part 3. They can be a good means of overcoming barriers between researchers and farmers and sometimes extensionists as well. For example, the case study contributed by Abedin and Haque (Section 3.1), illustrates an 'innovators workshop' in Bangladesh, which helped farmers to reveal their discoveries and knowledge and researchers to break down their prejudices. The workshop was attended by science-trained professional researchers and 30 extension officers who 'enthusiastically' learned from four farmers who described their inno-vations in potato farming. Discussing farmers' innovations in such group meetings is one way to bridge gaps and overcome biased views.

'Visual-aided dialogue' is another good method for researchers with farmers and for farmers with researchers. This involves discussing items which are in front of the participants. Insect pests can be shown to farmers to elicit comments on pest control, or farmers can display and discuss the varieties of seeds in their stores and discuss their preferences and reasons for them. Using such visual aids has been found useful at the beginning of a research process, to break the ice and establish rapport, to stimulate interest, curiosity and discussion and to enable researchers to learn from farmers.

Another method is what Anil Gupta calls manual discriminant analysis. This can be used in group discussions among farmers and researchers about farming practices and constraints. The basic concept is to compare and contrast. Farmers are asked to describe their management practices, for example for tillage, cropping patterns, input use, or techniques for maintaining soil fertility. Data are collected to find out the range of plotwise practices. The researcher then contacts the farmers whose practices are at the two contrasted ends of the distribution. Each group is asked to explain the behaviour of the other. For instance, farmers using the highest seed rates are asked about farmers with the lowest seed rates, and vice versa. This calibrates the frame of reference and makes clearer to farmers the distinctiveness of their own behaviour. Only after hypo-thesizing why the others behave differently is the group asked to explain the reasons for its own practices.

Although one of the objects of such discussions is to separate out the role of class from that of ecology, Gupta suggests that the researcher should avoid asking leading questions, or using the categories of rich or poor farmers when asking questions. If practices differ on the basis of class, it will become apparent from the data (Gupta 1987a).

Matrices to list who will gain and who will lose from alternative interventions or technologies are a simple check that outsiders can use. One form of this is what Rocheleau calls 'multiple-user analysis'. In this, multiple users of the same resource are identified. They may be women and men in households and farms, or different social groups. The analysis

44

can be used to modify action to make it more equitable especially for the disadvantaged and those who might be left out.

Extensionists and farmers

Recognition of farmers' knowledge and innovative capacity does not necessarily mean that they do not need extension services. Rather, it points to needs to improve the interaction between extensionists and local people to reverse and balance conventional 'top-down' communication and to overcome gaps and miscommunication. Some of the contributors to this volume describe how to improve such interactions.

For example, Suriya Smutkupt (1987) sees the role of extension officers as 'facilitators', promoting interaction between farmers and thus encouraging farmer-to-farmer extension. He describes the application of this idea to the dissemination in Thailand of a cropping system in which peanuts are grown after rice. The facilitator concept is useful. Farmer-to-farmer interactions are as important for innovation and development as farmer-researcher interactions.

Another example, described by Noel Chavangi and Agnes Ngugi (1987), is tree-planting in Kenya, to which extensionists, researchers and farmers all contributed. In the beginning of this programme, four new quick-growing trees suitable for fuelwood production were introduced, including *leucaena*. Each of 28 farmers were given 15 to 20 seedlings of three of the species and 50 seeds of one of them. The programme staff assumed that the farm people could be left to grow the new tree species *without* advice from extension workers, because they knew that farmers in the area had been planting trees for decades on their own. Therefore, all the farmers were told was that these were quick-growing fuelwood species similar to *Sesbania sesban*, a tree they already knew.

This approach of providing minimal advice 'worked' to the extent that the farmers successfully established trees from most of the seeds and seedlings. However, harvesting the trees for firewood was sporadic because people did not know how big they would grow or whether they could be coppiced. It was clear, in retrospect, that information from extension workers was needed on these matters. The staff also realized later that they had not adequately understood important social issues pertaining to tree planting, in that trees on a man's farm are seen as his property, but responsibility for gathering firewood falls on the women. Therefore, trees grown especially for firewood were perceived differently by each gender group. Where men expressed interest in planting more trees of these new species, it was often with the intention of using them to produce poles for construction and fencing.

In the next phase of the programme, more extension work was done to explain the growth characteristics of the trees and harvesting techniques, pointing out that some species could be coppiced for fuel within two years. The fuel-wood issue and its social implications were also discussed.

This project raises an important point about the nature of innovation. Farmers certainly possess relevant knowledge and are keen observers of trees, herbaceous plants and soils and they often carry out experiments, but knowledge, observation and experiment are not the only roots of innovation. Interaction between people with different kinds of knowledge and different areas of experience can be a necessary stimulus. A strategy of minimal extension work leaving maximum scope for farmer initiatives may give insufficient stimulus as well as insufficient information about an unfamiliar species.

However, the open and flexible approach adopted in this tree-planting project makes a welcome change from the old habit of offering new technology as a 'package' of materials, tools and procedures which was not supposed to be modified, and which made stimulating interaction all but impossible. One example was a package of watershed management techniques introduced into a dryland farming area in India under British auspices. Describing the experience, Verma (1987) comments that in the absence of 'adequate contact of the committee members with the common farmers', the project staff failed to notice that the farmers, in fact, disapproved of the soil conservation measures they proposed to introduce. Only after a near-stalemate forced the project scientists to be more responsive to the farmers did a compromise programme begin to make headway.

Another case of a similar lesson was discussed by Bashir Jama, who describes a Kenya government agroforestry programme in the Kilifi District. In this situation, researchers and extensionists (from both agriculture and forestry disciplines) began agroforestry and afforestation projects with a 'top-down approach', with little farmer participation; this involved mainly on-station trials and demonstrations and attempts to induce farmers to adopt a technical package of *leucaena* alley-cropping which included management recommendations. The researchers' main interest was to promote a way of using trees as nutrient-pumps and for nitrogen-fixation. After the initial stage, it was realized that this was inappropriate, because it did not meet farmers' immediate priorities. The farmers rejected a 'rigid' package; they instead modified the alley-cropping according to their knowledge, needs and own inventiveness.

After the staff recognized the problems with their approach and the importance of the farmers' initiatives, they changed the researcher-extension-farmer relationships and adopted a 'bottom-up' approach which emphasized learning from farmers and giving priority to local choices and ideas. As a result, farmers adopted *leucaena* not just to help soil fertility, but mainly for the leaves for livestock fodder, which was scarce and valued in the area. As a 'spin-off' result, they developed milk production from the improved livestock fed on the new fodder and farmers' earnings from milk sales then led them to establish a dairy development project. The resulting farmyard manure also increased yields of food crops substantially. Thus, the farmers' initial adoption of agroforestry ideas, along with flexible interaction with supportive extensionists, led to a chain of useful unexpected events and the project developed its own momentum and sustainability, based largely on farmers' initiatives.

Experiences of this sort demonstrate why there is need for a more open, interactive approach in which a 'basket' of technologies are offered (Maurya, Section 1.2) instead of complete packages, with a range of alternatives from which farmers can choose. The old idea of a 'transfer of technology' to the farmer from some island of expertise is thus being displaced by something more like a technology exchange, with benefits on both sides.

Local women, men and specialists

Social and cultural interactions between members of households and between specialized groups in society also help in understanding and local innovation. Complex social and cultural relationships and norms affect the use and ownership of resources, how farming operations are undertaken and how new ideas and technologies are perceived. Many family members may also be involved in processes of innovation. Within the family, male-female interactions influence innovation, notably with regard to crop varieties and domestication from the wild. The different roles of men and women in farming families have attracted attention, as described in Section 1.4 where men claim to 'own' and control a resource such as trees, but women have responsibility for using that resource eg, as fuel. Kean (1987) and Rocheleau (1987a) independently report work in an area of Zambia where innovations adopted by women both on the *chitemene* plots of shifting cultivation and in home gardens, are strongly influenced by what they can control independently of men. Rocheleau uses a schematic map (figure 1.3) to show how roles vary on different parts of the farm. She then goes on to show how research priorities for agroforestry drawn up by scientists were both altered and enriched by more detailed investigation of these arrangements through interaction with both women and men farmers.

In many communities, women not only control crop processing (figure 1.3) but also look after the family grain store. It is thus often the women who control the selection and storage of seed and its separation from food grain. So when there is talk about how farm people select and improve crop varieties, there is often need to investigate where the key decisions are made. If a substantial proportion of the crop is sold, it will usually be the man in the family who decides what is kept for seed; but where food is kept for home consumption, women may have a bigger influence and may use different criteria for selection. It is likely that in many societies, both men and women contribute to the selection process. The home garden is also important for selecting crop varieties and domesticating wild plants. All kinds of plants may be grown there which farming families do not want to grow on a large scale in their fields, but which they wish to have for the sake of variety in their diet, to produce a reserve stock of seed or just for experiment and to observe them. Potato or cassava land-races which people suspect may be more hardy or resistant than new varieties may be grown in gardens, where extension workers do not see them, so home gardens are an important means of maintaining genetic diversity. As with grain stores, women have responsibility for these gardens in

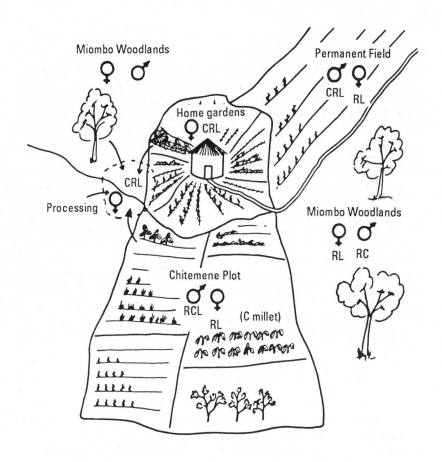

Figure 1.3: *The* chitemene *system in north-east Zambia, including new practices observed near Misamfu. Women control the millet crop (one of several in the intercrop rotation) on the* chitemene *plot. Other aspects of gender roles are denoted by:*
    C 'control';
    R 'responsibility';
    L 'labour'.
(after Rocheleau 1987a)

many societies and control most (but not all) of what goes on in them (figure 1.3).

New practices and stimuli for innovation also come from other interactions involving the market and local specialists. Traders bring new products – machines, tools, fertilizers, pesticides and seeds. Artisans such as blacksmiths and mechanics modify some of these products or develop their own. In India they developed small threshers to replace the more elaborate ones which were initially supplied. Farmer-artisan interaction can probably be credited with the development of a specialized blade-hoe in Haryana which helps to save time and conserve moisture. Or again, villages often contain individuals who are experts on trees, herbs or spices or who practice herbal medicine and who also experiment and innovate. No less than farmers, other specialists are subject to outside influences and also innovate.

Local knowledge and new technology interactions: costs and benefits

The adaptation of local knowledge to particular environmental and cultural 'niches' has value for the livelihood of a family or community, but it can be a limitation where socio-economic pressures are rapidly emerging. Disruption by environmental catastrophe or war can make some knowledge irrelevant. In such cases, new technologies may be especially needed. Farrington and Martin (1987:21) comment that in situations 'where ITK has been deprived of the social context necessary for its implementation, it may retain a certain potential for interaction with formal knowledge systems, but external intervention will be necessary . . . to help re-establish organizational patterns and to restore confidence . . . in traditional knowledge systems'. As an example, Farrington and Martin quote a soil and water conservation programme in the Yatenga region of Burkina Faso. Farmers were leaving the land as soil erosion destroyed its productivity, but a 'sensitively managed external intervention' (from Oxfam) and a new technique for surveying contours led to a renewal of indigenous soil and water conservation methods and significant improvements in crop production (Wright and Bonkougou, 1987; Pacey and Cullis, 1986:166).

While new technologies developed with the aid of conventional science sometimes present opportunities for rural people, they also pose threats. For arable farming, the classic example of the previous generation was the wheat and rice technology of the 'green revolution' resulted in both promises and problems. For pastoralists, improvements in human health, veterinary services and water supplies have similarly held out prospects for improvement while at the same time leading to environmental problems and human tragedies. Writing about African pastoralists, Barrow (1987) comments that pastoral people have rich traditional knowledge of land use but because changes in veterinary medicine and other fields have been so rapid, it has not been possible to adjust to the pace of innovation. This has led to the breakdown of many traditional controls on land use and has contributed to the degradation of land. Responding to this situation, many 'experts' on range management feel they have the answer to the problems

49

of particular areas based on technical parameters such as rainfall, soil and crops. They forget that development is about people and that necessary technical measures need to spring from interaction among people.

Technological developments can be seen not only in terms of their material impact on disadvantaged people, but also in relation to their psychological effect. If imported technology seems all-powerful, people may feel that their own efforts to improve land and grow better crops are futile. The formal schooling to which their children are exposed often tends to reinforce this demoralizing effect by disregarding local needs so that educated youth is no longer equipped to live in the rural environment. In pastoral areas of Kenya, children have already learned a lot about trees from their parents before they start school, but rather than building on this knowledge, the teaching tends to marginalize it (Barrow, 1987). Schooling may also undermine traditional rural knowledge by implying that much of it is mere superstition. Such influences go a long way to explain why farmers lack confidence in what they are doing, hiding traditional potato varieties from extension experts, for example. Thrupp (1987) mentions encounters with farmers in Costa Rica who showed embarrassment when asked about their own ideas and practices, explaining apologetically in one case: 'I don't practice the modern techniques that the experts say I should.'

For the current generation, innovations arising from biotechnology seem likely to pose an even more potent mixture of benefits and costs. Some believe there is the prospect of a new agricultural 'revolution' based on new seed varieties, new material for vegetative reproduction, and new vaccines for animals. A vast tide of fresh material is coming and seed companies will introduce it rather than research stations. This raises several potential dangers: for example, farmers may not be consulted about their needs and potential uses of biotechnologies. Furthermore indigenous genetic seed resources may become extracted and exploited by external interests or displaced by the spread of dominant new strains. Some analysts have warned that new genetic material could solely benefit resource-rich farmers and agroindustry, exacerbating social inequities and neglecting the needs of the poor. There is some hope, though, in the possibility that the promoters of biotechnology will see their opportunity for the largest profits in areas not affected by the 'green revolution'. In that case, there may be a chance of steering the new technology in the direction of reducing disadvantage.[9]

Notes

1 Sentence and reference added by editor.
2 Based on pages 8–13 in Maurya and Bottrall (1987).
3 Editor's note: Similar lessons have emerged in other AF projects which will be discussed in later sections (eg, 1.7 and 2.4).
4 This section is based on Gupta (1987b), pages 1–3.

5 Based on Gupta (1987a), pages 43–45.
6 Based on Gupta (1987a), pages 26–32.
7 Based on Gupta (1987a), pages 2, 12–14, 31–2.
8 Based on discussions in the ITK study group and comments by Anil Gupta, Ed Barrow, Roland Bunch, Calestous Juma, and Lori Ann Thrupp.
9 This paragraph is a near-verbatim note of an oral comment by Jacqueline Ashby.

PART 2

Farmers' agendas first

Introduction

Most professionals assume they know what farmers want and need but are often wrong. Not knowing farmers' priorities and not putting farmers' agendas first mean that professionals are likely to address the wrong problems in their research. Conversely, identifying farmers' priorities and helping farmers meet them leads to innovations which are adopted.

Part 2 describes approaches and experiences which have been developed to enable farmers to take more part in analysis and in identifying priorities. Roland Bunch presents a dynamic view of agriculture and describes tested methods for strengthening farmers' experiments, helping them adapt better in changing conditions, and empowering them to develop their own agriculture. Louk Box outlines how in the Dominican Republic crop ethnohistories, local networks and interest groups, and farmer experimentation together contributed to problem identification for adaptive research. SB Mathema and Daniel Galt describe the 'group trek' in Nepal which brings scientists closer to farmers and to each other in a shared field exercise to identify location-specific research problems. Corazon Lamug outlines practices developed in the Philippine uplands for community appraisal, community organization and process documentation using a battery of methods which bring farmers and scientists together. Gordon Conway presents diagrams that can be used with and by farmers for local analysis, and Anil Gupta and others expound approaches with mapping, and aerial photographs. Finally, Clive Lightfoot and his colleagues describe a process in the Philippines in which scientists helped farmers to conduct their own analysis, using systems diagramming to specify their agenda and then devising and conducting their own experimental treatments with scientists' support.

To put farmers' agendas first requires diagnosis in which farmers take part in analysis and in which sensitive researchers respect farmers as people, professionals, and colleagues.

2.1   Encouraging farmers' experiments

ROLAND BUNCH

Agricultural development goals and farmers' innovations

Two basic, nearly universal assumptions have, until the last few years, drastically reduced the effectiveness of agricultural development efforts around the world.

The first is that the basic goals of agricultural programmes should be to teach small farmers a set of innovations that will increase an area's productivity, and that, having adopted these practices, the people will continue indefinitely to farm at the new, higher level of productivity. This assumption is, in most cases, simply mistaken. A productive agriculture requires a constantly changing mix of techniques and inputs. Seeds degenerate, insect pests spread and develop resistance, market prices fluctuate, new inputs appear and old ones become expensive, agricultural and trading laws change, and temporarily successful technologies become less profitable as their spread forces market prices downward.

If it is true, then, that only a few innovations will ever be permanent, our possibilities for sucessful development may seem remote. What alternative do we have? The only practical one is to encourage a process by which people can develop their own agriculture. The goal of an agricultural programme, therefore, should be to train and motivate farmers to teach each other the innovations learned from programme staff and then to encourage them to improve on those innovations by themselves. Through a process of small-scale experimentation, farmers anywhere can develop and adapt new technologies that will carry their production on to steadily higher levels and by learning to become teachers of these new techno-logies, they can spread them throughout the programme area and beyond. Five years after the external staff have left, production levels, far from having dropped, should be higher and improved production more widespread.

Perhaps one of the reasons why new programmes have moved beyond this first assumption is the widespread (and paternalistic) belief in the second assumption that villagers are incapable of inventing, developing and adapting new technologies, thereby carrying on the agricultural development process by themselves. Evidence to the contrary not only exists quite widely in the literature, but is piling up at a rapid pace in the records of programmes supported by agencies such as World Neighbors (WN) and which provide the evidence and experience which are reported on here (see also Bunch 1985, 1988). The basic model is that an outside agency works with farmers in an area for perhaps five to eight years and then is able to withdraw, leaving behind farmers with an enhanced capacity to innovate, experiment and adapt.

Of course, traditional agriculture itself can be seen as the long-term outcome of technologies developed by villagers or disseminated by vill-agers. Small farmers on several continents have proven themselves capable of developing new kinds of technology in such categories as soil conser-vation, plant spacings and populations, intercropping, non-toxic pest and disease control, uses of native species, tools and labour-saving techniques. Nevertheless, here I will list examples that have been observed relatively recently in the areas where modern agricultural development programmes have consciously stimulated villager-initiated and villager-managed research.

To begin with *soil conservation*, farmers in central America have been observed to experiment with swaths of compost along the edge of contour

56

ditches and to try new uses of Napier grass or retained bands of natural vegetation to check erosion.

Innovations concerning *plant spacings and populations* are far too numerous to list individually. They include planting crop seeds in hills or mounds with different spacings and numbers of seeds in each one, planting in double rows and myriad adjustments of traditional intercropping.

Possibilities for intercropping can be illustrated by one of a vast number of innovations introduced by farmers. In the San Martin Programme in Guatemala, farmers just learning to grow groundnuts tried interplanting them with beans (a traditional crop). The beans, which matured more quickly, were harvested before the groundnuts needed all the space. Production was increased 50 per cent over separate plantings.

Non-toxic pest control was used in Guatemala when wild rabbits were wiping out programme-introduced soybeans. One day a local farmer smelled a horrible odour as he was walking by a drug store. It was iodine. He bought a pound, mixed it with water, and spread the solution around the borders of his soybean field. The rabbit problem was eliminated. In Chapare, Bolivia, where farmers were using the traditional jungle slash-and-burn agriculture, most of them were losing half their rice to an insect pest. Extensionists were recommending various insecticides, all of them expensive, toxic and rarely available. One day the extensionists came across a farmer whose rice was surprisingly undamaged. For three years, this farmer had kept his field free of the insect by clearing the jungle in such a way that the wind could circulate well, burning the host weeds thoroughly and planting on a certain date.

Uses of native species can be illustrated from El Rosario in Honduras, where farmers found that a local legume (velvetbean) which was being pushed by the programme as an intercropped green manure could be used as a mulch instead, after the beans had been harvested for human consumption. Another example from Honduras concerns farmers who have found four species of local grasses that can be substituted for the Napier grass now widely used for soil conservation in the area. One gives less shade to nearby plants and is good for intensive vegetable plots, while another has faster growth.

One of many *labour-saving techniques* was invented in Chiapas, Mexico, when farmers found velvetbeans growing wild in the nearby jungle and noticed that they shaded out all other weeds. They planted the beans with their corn and in traditional jungle clearings, but with the addition of chemical fertilizer, and were harvesting 4 tons/ha in the same fields year after year without the benefit of either a crop rotation or periods of fallow. A farmer in the World Neighbors Highlands Programme area of Peru who was managing a eucalyptus nursery at 3,800 m above sea level, was having to cover the seedbeds with plastic every night in order to protect them from heavy frosts. Later, he discovered that when he located the beds in a small forest of eucalyptus that he already had growing, the plastic cover was unnecessary.

All sorts of *tools* have been developed in various programme areas, often in response to new crops or operations introduced by development programmes. These include various sizes and shapes of hoes; long-handled pruning instruments; hand cultivators; maize-husking and maize-shelling tools; techniques for irrigation water distribution; and home-made modifications of various commercial tools, such as hand-sprayers.

The evidence of small-farmer spread of technologies is also significant. Almost all traditional agriculture is a result of this spontaneous spread of innovation from one farmer to another, from one village to another, and even clear across continents. For instance, at least two of the major crops in West Africa today, maize and cassava, were spread across an entire continent in less than 450 years, with no development programmes or agricultural extensionists anywhere in sight.

But apart from the spread of traditional technologies, there has been considerable evidence of the spread of programme-introduced technologies far beyond where they were introduced and long after the programmes have left. For instance, in the WN Programme in Cebu, Philippines, soil onservation techniques had spread at least 5 km beyond a river that no programme personnel had ever crossed within four years of the programme's inception. And farmers within the programme area were growing at least eight different varieties of vegetables, several of them on a commercial scale, that they had learned to grow from other farmers. In Guatemala, WN/Oxfam Programmes at Poaquil and San Martin plus the WN Quetzal Cooperative extension work resulted in the adoption of contour ditches and Napier grass barriers five years later and as far as 15 kms from the area where the programmes had introduced them. Furthermore, farmers within the programme areas were growing strawberries, cauliflower and broccoli, which they had learned to grow from other small farmers outside the area.

Why promote and teach small-scale experimentation?

Most of the innovations above were developed by small and often illiterate farmers, who were encouraged to be continually investigating new ideas and new technologies. They were motivated to do so by the programmes' careful selection of the initial innovations that were tried out, so that rapid and recognizable success made them aware of the significant practical advantages of experimenting.

The resulting self-sustainability of development is perhaps the most important reason for using villager experimentation in all agricultural programmes, but there are also other advantages to be reaped from teaching small-scale experimentation.

First, experimenting on a small rather than large scale reduces the level of risk, whether from crop failures due to adverse weather, inappropriateness of technology to farmers' conditions, temporary unavailability of inputs or inadequate understanding of the technology. If large-scale experiments fail, large losses of crops, animals or capital investment can result in family tragedy. With small-scale experiments, much less is at stake.

Second, a farmer can learn much more this way than by experimenting

58

with his entire crop. If a farmer makes a change in his entire crop, he can try only one technology or package of technologies each year. If, however, he devotes just 1/20 ha to small-scale experiments, he can do as many as 10 experiments occupying 50 square metres each in a year. Many farmers have already proven themselves willing to do this (Bunch 1985).

Third, a farmer who makes a change in his or her entire crop has no way of comparing the changed technology with the traditional. Thus, he or she may blame problems of weather or insect pests on the new technology, rather than their true source, thereby abandoning a good technology needlessly. On the other hand, a farmer who tries out new technology on a small-scale has the rest of his land as a natural control plot.

Small-scale experimentation also helps the extensionist preserve his credibility and prestige, two assets frequently lacking in agricultural extension work. If an extensionist (who is always preferably also a villager) needs to convince the farmer to try out an innovation on a small plot, he can present the idea as just an idea worth trying, rather than getting involved in the promotion of the 'sure-fire successes'. Thus, if the technology does fail, the farmer is not likely to feel that the extensionist cheated him or lied to him. Furthermore, the feelings of hurt, frustration or anger that are easily engendered by the loss of an entire year's crop can largely be avoided if the failure occurred in only a small fraction of the farmer's fields. Thus the extensionist's job is made easier and more fruitful. Extensionists are protected against a loss of credibility and friendship among the villagers and the technology can reach more farmers because they risk less in trying it out.

Further advantages accrue to such programmes of small-scale experimentation. First, they can reach the poorest farmers because the small scale reduces the cost of technology adoption. Second, should a loan service be provided, values of loans are reduced, more farmers can be assisted, severe indebtedness is avoided and better repayment rates are permitted. Third, as farmers increase experimentation, the programme staff will begin to learn more about their own technology, as well as learn of farmers' new appropriate technologies adapted to local conditions. This process increases villagers' dignity, converts 'extensionism' into 'communication', and increases the quality and range of the appropriate technologies available for teaching.

How to teach small-scale experimentation

Villagers at first need only learn to:

- measure off several plots of land or separate out two or three animals;
- plan their experiments so that only one production factor varies between each two plots or groups of animals;
- weigh or measure the results; and
- write down and add up all the expenses and income of the experiments as well as the controls.

Farmers of nearly every educational level have proven themselves capable of doing this. Even illiterate farmers have been taught to read and write numbers and then use mimeographed sheets with drawings that depict the various cultural practices and inputs. In time, as farmers become more eager to learn and able to understand more, they can keep more exact accounts and use more complex, scientific experimental designs, although the value of such complications are often overestimated by professionals.

During a programme's last year or two, programme outsiders (the personnel not consisting of villagers from the area) should gradually phase themselves out of the training function. At the same time, the programme should put its villager leaders in contact with permanent sources of technology in the area: experiment stations, commercial agents, other innovative villagers, helpful radio stations, professional employees of nearby local institutions and sympathetic agronomists. This should help prepare the farmers for the time after the programme staff have left the area when the farmers' ideas for experimenting will come from their own natural inventiveness and also from the above outside sources.

As the farmers try out more and more innovations, the village leaders should learn to share their new-found information with each other. Increasingly, villager extensionists and leaders should use the training classes they attend to show and report to each other the results of their own experiments.

In time, they should gradually organize themselves to coordinate their experimentation. For instance, one group of farmers could experiment with five new groundnut varieties while another group takes charge of comparing various plant populations for millet.

Using a standard mimeographed 'experiment report sheet', these groups of farmers can then report their findings to the larger group, representing perhaps 30 or 40 villages. By thus organizing what is in effect their own experiment station, stretched out across the programme area, the villagers can develop locally adapted technology. By systematically communicating with each other, they can spread this technology across the entire area. The organizational framework in which this process occurs may be a formal institution or just groups of farmers who have come to know each other.

Even if the traditional experiment stations do begin to develop more technology appropriate to small farmers, we can hardly expect them to develop the technology needed for the uniquely different farming systems evolved by farmers in the area of a truly successful programme. Therefore, a relatively self-contained system of developing and disseminating new technology such as that described above becomes essential if farmers are to continually innovate in a self-sustaining manner.

Thus meetings held by the programme will be increasingly taken up by the villagers reporting on experiments, or arranging new experiments among themselves. Programme staff who are outsiders to the village are thus gradually phased out of teaching and a shift to locally taught and organized classes will be made. The dialogue once carried on between the external staff and the villagers will have become a dialogue between groups of villagers themselves.

Programmes that have carried this process through to its final stage are, as yet, rare. Nevertheless, successes have been observed and evidence such as that cited above is accumulating to show that farmers have become increasingly capable of carrying on what is both the least often accomplished and the most important goal of agricultural improvement work: the never-ending process of people experimenting and innovating to develop their own agriculture (Bunch 1985).

2.2 Virgilio's theorem: a method for adaptive agricultural research

LOUK BOX

Building interfaces

When we had just met, Virgilio stood up and said: 'Lucas, I understand you want to know. You are a scientist and you want to know. But there is only one way to know what I know about cassava. Speak with me; don't speak to me like others did. Ask me about my life and I will tell you about cassava.'

It happened at the outset of my research in the Dominican Republic, in 1980. The remark struck me because it reminded me of comments in the literature. Farmers get tired of being talked to when they should be talked with and here it happened: the man who sweeps the street in the small provincial town of Moncion, telling me the same things social scientists repeat time and again. I had been sent to consult Virgilio by a well respected cassava cultivator in the neighbourhood who told me: 'that man knows more than anybody else; he is a wise man.' I spoke at length with him and it appeared that although he had benefited from only three years of formal education, he could write down much of his information. Although he was among the poorest of the village, he was duly respected for his knowledge about cassava. Although his social position was marginal in many respects, he was part of an invisible network of those who experiment with cassava.

Later, when the research had come in full swing, I was often reminded of Virgilio. His words remained true, and I had to force myself and my colleagues or assistants to use this notion of 'speaking with' instead of 'speaking to'. We took his advice to heart: asked about lives and heard about cassava. We reconstructed the history of cassava cultivation and so got an idea of why people had changed from one variety to another, or had opted for another cultivation practice. In so doing we designed trials together; we redesigned the experiments the farmers had been doing themselves to make them more reliable to the scientists we worked with and we redesigned the scientific experiments so they could respond to local conditions. We called this mutual adaptation and the experiments, adaptive trials.

This is the story of the research we did between 1980 and 1985 in the Dominican Republic. It was based on the notion that cultivators experiment and that scientists ought to be aware of this. At the time, little credence was paid to these notions. I recall my colleagues' wisecracks in Wageningen about 'experimenting farmers' and one of them pointedly asked me when the research proposal was considered: 'Mr Box, what *is* the scientific status of your "experimenting farmers"? Please show me one article which makes the case.' I could at that time point to one article (and only one: Johnson, 1972) and to one historic fact (the contribution of Dutch potato growers to the development of new cultivars). It did not impress, but sufficed as a minimal answer. Now, things have changed. Richards' (1985) book is common reading, Farming Systems Research has made its point and the respectful attitude prevalent among social anthropologists is making its way into international agricultural research centres (Rhoades 1984). Experimenting farmers are in. The question is: how do we keep them in?

This paper argues in favour of three techniques which are complementary to standard agricultural research practices. In essence, they were already mentioned in Virgilio's story and they are:

• *historic*: reconstructing cultivator biographies with respect to a particular crop, thus learning about the discontinuities in crop cultivation and the experiments done to adapt to changing circumstances or improve available technology. I shall call this technique *biographical analysis*;
• *agronomic*: translating cultivator experiments into scientific designs, plus adapting scientific trials to local conditions. I shall call this *adaptive trials*;
• *sociological*: transforming local knowledge about a crop into more general statements and articulating local knowledge networks with more general ones. This is done by constructing interfaces (a term inspired by Long's work (1986)) between existing networks, and I will call it *knowledge network transformation.*

My suggestion is that social scientists can provide effective and efficient complementary techniques to standard agricultural research wherever social distance between cultivators and technicians is great. Social distance increases when formal research and extension procedures are implemented without adequate regard for cultivator knowledge. Social distance can be reduced by linking networks through interface development. Interfaces are the points where networks meet, as in the case of the cultivator who exchanges information with the extensionist, the researcher or the trader. Although this paper is discussing research done in the Dominican Republic, preliminary findings for the Netherlands indicate comparable results.

Early steps in the research

The Adaptive Agricultural Research programme in the Dominican Republic ran from 1981 till 1985 (although I started preliminary work in 1979) and worked on two crops, cassava and rice. The account which follows is mainly about cassava. The research team varied over time, but mainly consisted of two sociologists (one for each crop), two agronomists (again, one for each crop), an extensionist, and an anthropologist working on women-related issues. A varying number of Dominican and Dutch research assistants did essential tasks connected with adaptive trials, case studies and interviewing.

We chose to work in a research institute, the Centro de Desarrollo Agropecuario Zona Norte, because it provided close contacts with the people who were making day-to-day decisions regarding research priorities. So we opted *not* to work in a university setting, in a farmers' organization or in the national extension service, although all these were suggested to us at one time or another.

In the research institute, we were the odd ones out in many respects – insisting on farmers' knowledge, where 'they' stressed scientific expertise; attempting to involve extension from the beginning, against the conventional wisdom that extension comes at the end; defining research priorities through agrosociological studies, instead of following the normal procedure of waiting for the Ministry to decide on them. In time the team developed a network going both into the roots of farmer organizations and into the upper branches of bureaucracy.

We first identified major interest groups in the research arena. These were:

- the researchers in research institutes, universities and elsewhere;
- the cultivators: marginal producers, cultivating a crop on marginal lands, to be sold among the urban (and rural) poor. Cultivators were organized to a certain extent in associations under the umbrella of regional federations. Differences between regions were large, especially between the prime sweet cassava region in the rich Cibao Valley, and the bitter cassava area in the hill country called the Sierra;
- the extensionists working in these areas: relatively young, inexperienced highschool or college graduates with many tasks and few resources;
- the agrobureaucrats working on regional or national levels; experienced officials with national or international contacts, sensitive to the political climate of the day;
- last but not least the private sector: varying from the small businessman buying the crop from local producers to the owner of the big concern with multiple linkages to international agricultural research and agribusiness.

For cassava, each of these sectors was diffuse in its internal organization (the contrary was true for rice). Cassava cultivation, research, extension, development policy and business are characterized by loosely structured networks, ill-articulated both within and among sectors. They were a starting point for our research. We could not count on articulated knowledge networks. If we were interested in knowing more about the crop we would have to identify and then develop the networks ourselves.

Our main worry in the beginning was to do work which our prime partners considered important. In the case of the cassava research this meant identifying researchers' priorities, speaking to agrobureaucrats to hear their views, and synthesizing existing studies. This resulted in a first study (Castellanos and Box 1981) on the state of cassava research in the country, characterized as long-standing but discontinuous, promising but with few concrete results and not linked to cultivator pressure groups. Through this study we came to know most of the researchers and established a network which proved essential in the years to come.

Right from the start we attempted to associate with farmer organizations in the prime growing areas. In fact we defined our research areas in terms of these organizations. The Valley area became identified with the Small farmer Federation of Moca; the Hill area with the Federation of Moncion. This meant that all our work was channelled by these organizations, that geographical boundaries were defined through organizational limits and that most contacts were established through them.

There were two reasons for doing this: first, confidence, which is the basis of all sound social science research and second, effectiveness and the belief that if this type of research is to be institutionalized it has to be through interest organizations like these Federations. Probably I was influenced by the Dutch example, where farmers are hyperorganized into solid networks and interest groups forging strong links between the worlds of farming, research and extension. In other words, we wanted to create an interface between our world of research and their world of cassava cultivation.

In practice this meant frequent informal meetings with Federation officials, seminars to transmit research results to participating cultivators and to Federation officials and involvement of Federations in the testing of proposed technology. Other interest groups were also involved, particularly through yearly seminars. We thus tried to strengthen informal networks, increase knowledge exchange and enlarge the interface along which such exchange could take place.

Gradually we came to know many cassava cultivators and were able to select key informants. For each of the two areas we decided on about twenty informants, selected on three criteria:

- respect among peers, indicated by replies to questions: Who do you go to if you want to know something about cassava or you need new planting material?
- experimentation, as indicated for example by the maintenance of a collection of cassava varieties to check yields under varying circumstances;
- capacity to verbalize results and willingness to collaborate on the design and execution of adaptive trials.

Biographical analysis for problem identification

Case studies were compiled for all informants by a sociologist working with an agronomist – the smallest nucleus in our team. This was the heart of our research and was based on a biographical analysis with regard to cassava cultivation. The key question was: What was the oldest cassava variety you remember being grown when you were young and what was it like?

This question forced the informant to try and remember as much as he could about particular 'classical' varieties – and I say 'he' because all informants happened to be men, in contrast to the situation faced by Fresco (1986) in Zaire, where cassava cultivation is almost exclusively done by women. The question was not threatening and confirmed the informant in his role. It provided us with two types of information: names of varieties and criteria for distinguishing between them. On the basis of the interviews we developed a listing of all names or labels and by cross checking tried to find synonyms (different labels referring to one variety) and homonyms (same label but referring to different varieties).

Moreover, the logic of classification appeared. Reference was made to plant or tuber size, leaf colour, trunk or tuber colour, texture, bitterness or poison danger and ecological conditions fostering or hampering growth. Gradually, the classification criteria became evident and so did the logic linking them.

The information on classical varieties was then contrasted with varieties having appeared subsequently. This provided us with a chance to explore the reasons for changing from one variety to another. Sometimes these reasons had to do with passing fashions; most of the time they dealt with a better adaptation to local (and changing) conditions. Cultivators could refer to declining soil fertility, changing soil humidity, differing crop uses or changing market preferences. In an hour or so, we would follow the whole 'biography' of a cultivator and his cultivars, reminded of Virgilio's theorem. The interview was then completed by asking some additional background questions on cropping patterns, household structure and attitudes.

The real moment of truth came when we would go through the whole conversation and determine the discontinuities in cassava cultivation: changes in varieties, in cultural practices or otherwise. At each such rupture the historic context was reviewed to make sure we understood it in the same terms. Informant and interviewers then analysed the biography and reconstructed history. This was one of the most exciting moments: many informants started looking at their own history as cassava cultivators in a different way. Of course they *knew* all along what they said; they just had not seen it fully in these terms. In Giddens' (1979) analysis they had been using the rules of cassava production and reproduction, but were not quite aware of them. The interview could therefore provide new insights to both parties, making the effort worthwhile to the cultivator as well.

Throughout the interview we did not ask for problems. My experience in previous research was that cultivators would give standard answers when asked for problems: low prices, exploitation by middlemen, poor roads or no access to credit. All of these problems may be real, but are always mentioned and do not provide much new knowledge on which agricultural researchers can base their priorities. We therefore refrained from such questions but rather deduced problems from the synthesis made towards the end of the interview. Each change in technology was linked to a specific historical setting, like soil degeneration, market incorporation or government intervention. After comparing these settings for change, we could isolate a number of problem areas like soil quality, root rot and the need for short-cycle varieties.

Problem verification and adaptive trials

Problem identifications were then fed back to group meetings, in which informants and others took part. Federation officials (cultivators themselves) could react to our findings and use the conclusions as they saw fit. In some meetings the researchers were provided with alternative explanations for changes, thus enlarging our model. The meetings were a necessary complement to the individual interviews and allowed for higher-level verification and generalization.

We used two other ways to verify the problems and other impressions gained from the informant case studies: a random sample survey among 247 cultivators and a survey among researchers, extensionists and others with regard to problem perception. In so doing, we obtained a unique description of the problems facing cassava cultivation in the Dominican Republic. Our impression about the different networks could then be put to a test.

We had observed the trials by both researchers and cultivators. It was evident that they were aimed at different objectives (Box 1987b), used different methods, and produced results which were not verifiable by the other party. This situation has been described for other crops (eg, potatoes, Rhoades and Booth 1982). We developed a procedure based on the notion of mutual adaptation. Cultivators were asked to adapt their trials to conditions allowing for statistical analysis. Researchers were asked to design trials adapted to productive conditions faced by cultivators.

With cassava, we tested erosion control by strip cropping; types of root rot under different conditions; and the effects of multiple cropping. Adaptive trials were now far more complex. Cultivators and researchers were required to make an extra effort in design, execution and analysis. Their logic differed so much that it was hard to come to terms. The role of the social scientist as a two-way translator became essential. When the trials were left to either party, they soon reverted to their old state.

The adaptive trials created working relationships between experimenting cultivators, agricultural researchers and social scientists. They required interaction between different networks. At first this was difficult. In the initial meetings, parties hardly listened to each other. An interesting example was the case of root rot. We found out that cultivators distinguished two types of root deterioration: root rot proper, and what they called *sancocho* (stew). *Sancocho* was associated with sudden rain after a period of great heat. The tubers were said to stew in the hot water which caused them to deteriorate.

Researchers did not know about *sancocho* and refused to believe in it. Only after physical proof and reference to a trustworthy publication from the international centre at which all of them were trained did the researchers accept this knowledge. At a closing seminar, where all networks were represented, a researcher acknowledged that there was

66

such a thing as *sancocho* and that it seemed important enough to investigate.

It is through such seminars that networks may link, and that social distance can be reduced. We also used our case studies, the surveys, the adaptive trials and occasional field visits for such purposes, but probably most effective was the notion of experimenting cultivators. Only when researchers could be shown that contemporary knowledge in cassava cultivation does not stem exclusively from their experiments, but from those performed by cultivators also, was a bridge made between these worlds which stand so far apart. As I have shown elsewhere (Box 1987b), extensionists cannot be entrusted exclusively with this job; they have been educated into one paradigm stressing the use of inputs. Simple feedback models, suggesting that extension will provide network-linkage, are therefore likely to fail.

Discussion and conclusion

Rapid rural surveys and Sondeos have their merits, provided one knows what questions to ask. If this is not the case, in-depth case studies need to be done. We found in both the cassava and the rice research that only through careful reconstruction of changes in crop cultivation patterns could questions be formulated (Box and Doorman 1985). Given the formidable gaps in knowledge about small-scale and rain-dependent agriculture in the tropics, scientists cannot assume that they already know the most relevant questions to ask cultivators. Our method amounts to a plea to start with cultivator experimentation, using this to define problems and later to graft on scientific experiments. In doing this, different networks may become better articulated, with better interaction between cultivators and scientists.

The social distance of scientists from farmers varies from crop to crop. Cassava cultivation in the Dominican Republic is a case of poor articulation between knowledge networks. With rice, we found greater articulation and better communication. Cassava could be representative for crops in marginal, rain-dependent agriculture. In the Dominican Republic, yields of these crops have declined over the last decade and research has not produced effective solutions.

If there is to be change, researchers need to be aware of the innovations cultivators are themselves introducing. Rather as Richards (1985) has argued for West African agriculture, the Dominican case shows that the conscious changes made by cultivators are a basis for further scientific investigation. Through biographical analysis, mutually adaptive trials and network articulation, interfaces are created, through which cultivator information and logic can be exchanged with interested researchers.

2.3 Appraisal by group trek

S B MATHEMA AND D L GALT

The samuhik bhraman process in Nepal

Toward the end of Nepal's Integrated Cereals Project (ICP) in 1984, it was realized that most of the impact at the farm level had occurred in one area and little or no appropriate technology had been developed for the hills. This is partly because the farming systems of the hills are more complicated and are integrated with livestock and agroforestry. To address these additional problems, it was proposed that a new USAID-funded project would follow ICP, and would use the farming systems research approach. Therefore, at the beginning of November 1985, Nepal government agencies, USAID and the Winrock International Institute for Agricultural Development agreed on the institutional framework for what became known as the Agricultural Research and Production Project (ARPP).

Working with Nepal government counterparts, ARPP evolved a rapid rural appraisal methodology which has been given the Nepali name *samuhik bhraman* 'a group trek'. The objective is to facilitate joint work by people from several agricultural disciplines and to interact efficiently with local farmers in a limited target area to determine problems and constraints affecting predominant crops, livestock and forestry patterns (Mathema et al, 1986).

There are two basic types of *samuhik bhraman*. The first is for initial exploration of the farming realities of a new site for farming systems research. The second is for following up farming systems in a given target area, and is not discussed here.

The first type, an 'initial' *samuhik bhraman*, is a form of rapid appraisal which culminates in the design of trials or other activities to resolve problems identified by farmers and researchers. The people taking part include commodity programme representatives, livestock and forestry experts, and extension workers. When they first assemble, there is first a brief orientation session, including a discussion of procedures for interviewing farmers. With regard to the latter, three approaches are distinguished: the key informant survey, individual farm household interviews and farmer group interviews.

In a key informant survey, a knowledgeable farmer (for example) describes the farming system followed by the majority of farmers in his or her village, but is not asked about his or her personal farming situation (Mathema and Van Der Veen 1978; Shrestha et al, 1987). By contrast, people do talk about their own farms in individual household interviews. These interviews are informal, and are conducted in a male or female farmer's own home, courtyard or field. Open-ended discussion takes place around a few 'key questions' and 'prompts'. Interviewers are urged to follow-up any interesting information which emerges and there is no time limit. The farmer group interviews may last two or three hours and may include a meal (which provides some incentive for the farmers to show interest). The purpose is generally to check that the information obtained about the local farming system stands up to group scrutiny.

Procedure in the field

The target area for research is normally three selected and contiguous wards (a Village Panchayat – the basic unit for local government in Nepal – always consists of nine wards, each with anything from 50 to 300 households). On arriving at such a site, two members of the research team – the 'site monitor' and a socioeconomist – will begin the key informant survey whilst the remaining researchers and extension workers form small groups to conduct individual farmer interviews. These groups, of two or three people each, are as multidisciplinary as possible, containing such mixes as: native Nepali speaker and expatriate staff member; social scientist and biologist; livestock specialist and rice breeder.

Key informants include the elected head of the Village Panchayat, the Pradhan Panch. He may be asked to suggest farmers who can be interviewed in this part of the survey, the aim being to interview six key farmers altogether, two in each ward. Other key informants may be ward chairmen, the district irrigation officer (if there is one) and any local representatives of the cooperative or the Agricultural Development Bank. All these people are asked to answer a short, formal questionnaire about population, the farming economy, and typical practices of local farmers. The key informant survey is thus an extremely parsimonious method of assessing ward-level conditions, resources, and practices (Mathema and Van Der Veen 1978).

Meanwhile, the other interview groups spread out across the target wards to conduct the individual farmer interviews. Each group completes one to three farmer interviews of between one and two hours each. Every attempt is made to interview both male and female farmers. If possible, interviews are conducted in the farmers' fields. However, the timing of the *samuhik bhraman* is usually during a period when the farmers of the Village Panchayat are not excessively busy and thus are often found at home.

Later in the afternoon or early evening, the technical group assembles to go over the information that has been gathered. This part of the exercise is very similar to the evening activity of the sondeo (Hildebrand, 1979; 1980; 1982; Ruano, 1982). One difference is that the information gathered from the key informants is also discussed and synthesized along with the individual farmer interviews. Key informant data are important in initial determination of Panchayat and Ward-level parameters (ie, proportion of various castes or ethnic groups; areas devoted to Panchayat or Ward protected forests; communal grazing areas and locations).

During this group discussion, each person in the group begins to feel more and more comfortable as gaps in individual knowledge bases are filled in and supplemented by the findings of others. At this time, the group also begins to make three master lists. The first is a list of the areas and items which need further clarification from farmers or key informants the

69

following day. The second is the beginning of the master list of farming systems and farmer's problems and constraints. The third is a list of tentative farm-level trials. Often, this third list will not be started until another day of interviews and interactions has taken place.

The next day, individual farmer interviews are continued, with the composition of the small interview groups being changed to lessen investigator bias. In addition, and depending upon the appointments made the previous day, one (or more) farmer group meetings may take place. At these group meetings, the technicians use the list of areas and items for further clarification (begun the night before) to help focus better on farmer-identified problems. Other issues raised by farmers are also noted for discussion that evening and for further consideration during the week.

During one initial *samuhik bhraman* (in Kotjahari), the wife of the chairman of Ward One also happened to be the representative from the Village Panchayat to the district Woman's Club. She called a meeting of women farmers for the afternoon of the second day (FSRDD, 1987). This meeting was attended mostly by women of the Ward and a few of the technicians. The Pradhan Panch, after giving his blessing to the meeting and to the concept of technicians working directly with the women of the Panchayat, walked over to another part of the school yard and met with the male farmers and the remaining technicians. Issues raised by both the women's and the men's groups were noted down and discussed later by the technician group.

Again on the evening of the second day, the technical group continues to synthesize and discuss information gathered from the area's farmers. At some point in the process, the group begins to write individual sections of the *samuhik bhraman* report. This phase consists of assembling the information gathered so far – on crops and cropping patterns, livestock, agroforestry and socio-economic factors – into draft sections written for the final report.

Much of the rest of the *samuhik bhraman* is spent in:

- farmer group meetings;
- beginning (or continuation of) individual report section write-ups;
- the review of these sections by the entire technical group; and/or
- the follow-up of any loose ends or contradictory findings (FSRDD, 1986; 1987).

Farmer meetings are held to reconfirm the information received in individual farmer interviews, as well as to refine it. Such meetings may be held on a Ward-by-Ward basis, or by dividing farmers along gender lines. At this time, farmers may be asked questions about improvements or technologies they may have tested in their own fields. Farmer responses are recorded and taken into consideration when trial design moves ahead (Mathema et al, 1986).

An example of following up a loose end and trying to reconcile contradictory findings is provided from the Kotjahari *samuhik bhraman* (FSRDD, 1987). Even though the group had heard much about community forests during their first three days in the Village Panchayat, the

70

details of each of the selected Ward's forests were still unclear. There were contradictions between key informant surveys and farmer interviews, between farmers' responses on different days and between individual interviews and the two group interviews. To gather first-hand information, a sub-group of technicians walked through one of the Ward's forests where undercover grazing of livestock is freely permitted and back through another which is protected from grazing for most of the year. This enabled the group to observe the extent of each Ward forest and the condition of each.

On the third afternoon or evening, the group achieves consensus on a *prioritized* list of farmer problems. They rank these problems and constraints from worst to least limiting. If time permits, the group begins trial design based on the prioritized list (FSRDD, 1986; 1987).

Prioritization of farmer problems by technicians is a dynamic process. After spending three days with the farmers of the selected wards, technicians have a pretty good 'feel' for many of the realities facing farmers. Disciplinary and commodity representatives are less strident and demanding in their suggestions for trials addressing their own disciplinary needs. Instead, the group is usually ready to prioritize based on two overriding criteria:

• the severity of the problem or constraint on the farm household, given the importance of the crop, animal or forestry component in the system; and

• the appropriateness for research on each selected topic of priority.

This second criterion means that high priorities given by farmers to 'rural electrification of the Panchayat' are beyond the research and extension mandate of the Ministry of Agriculture, while white grubs in maize is both a researchable problem and within the terms of reference of Ministry staff.

The fourth and final day consists of completing trial design, agreeing priorities for research trials, and, if time and situation permit, sharing summarized farmer problems and trial priorities with farmers in a final farmer group meeting (FSRDD, 1987; Mathema et al, 1986). If time or situation do not permit this final step, it is completed the following day in most cases.

The final farmer group meeting is not held if:

• the *samuhik bhraman* is evaluative in nature – that is, assessing on-going farming systems research;

• the *samuhik bhraman* is taking place on a *potential* research site which has not yet been finally selected as a location for farm-level research; or

• if the *samuhik bhraman* is used as a training exercise.

In the latter two cases, it is unwise to arouse farmer expectations for collaborative research if the area does not turn out to be chosen for farming systems research.

Sometimes it is possible to involve farmers informally in either the trial design or trial prioritization phases. For example, in Kotjahari, when farmers gathered around the technical group during final research trial prioritization, the group was able to solicit and receive farmer consensus

71

immediately on a couple of points which still troubled the technicians in reference to the white grub problem in maize (FSRDD, 1987).

Finally, the technicians trek back to their respective points of departure. The field report of the *samuhik bhraman* should be in draft form at this time. Follow-up trials are instituted either by the Farming Systems Research and Development Division, or by concerned disciplinary or commodity programmes with the collaboration of the site coordinator. Examples from the first *samuhik bhraman* in Naldung village panchayat include: winter wheat farmer field trials carried out and monitored by the site coordinator and site monitor; potato variety trials implemented by the National Potato Development Program and monitored by the site coordinator; a peach tree variety trial, implemented by the pomologist of the Horticulture Division and followed up by him and the site coordinator. But of course, not all proposed trials are carried out. That is one of the main reasons for drawing up priorities.

Conclusion

It is now widely accepted that the *samuhik bhraman* or group trek has several advantages over other types of rapid rural appraisal, at least in Nepal. These may be summarized as:

• researchers now have a vested interest in the technique because they enjoy the experience of combined trekking for the purpose of making farm-level research more relevant for the farmers;

• the trek is interdisciplinary and interdivisional. Participants may meet colleagues in other divisions of government departments for the first time on these occasions;

• the group trek places researchers and extension workers in physical contact with the reality of the farmers they are serving;

• the treks give an 'equal opportunity' to those who feel they should be involved in research decisions. The very fact of participation in the *samuhik bhraman* gives disciplinary and commodity representatives a voice in setting priorities for research activities. Everyone appreciates this, which encourages trust between divisions and departments;

• a group trek allows the joint setting of site-specific research priorities. Joint agreement thus replaces the piecemeal priority-setting process which otherwise occurs. The piecemeal process was one in which each site coordinator was collared at different times of the year by different experts or divisional officials and subjected to differing opinions as to what the farming system research should be;

• group treks make highly efficient use of scarce technical manpower. Rather than scheduling trips at random and at the whim of a separate discipline or community programme, two *samuhik bhramans* are sufficient, and replace numerous uncoordinated journeys;

• costs are low. However, there are problems about per diem expenses for the local staff which remain to be settled.

Finally, it must be remembered that the group trek process is new and evolutionary in Nepal and has been changing to meet farm-level realities.

However, the relevant government research divisions are extremely pleased with the results so far. For many of these reasons, the *samuhik bhraman* has become the backbone of diagnostic research within the Farming Systems Research Development Division (FSRDD, 1987).

2.4 Community appraisal among upland farmers

CORAZON B LAMUG

Upland forest agriculture in the Philippines

Upland farmers in the Philippines are those who cultivate crops on hilly, forested lands which are under the jurisdiction of the Bureau of Forest Development (BFD). Because traditional forestry has always given primacy to the production of trees, the human occupants of the forest were never given any attention except to fault them for their destructive cultivation practices, notably shifting cultivation with its requirement for cutting and burning forest vegetation. Such cultivation was thus perceived as mainly responsible for the destruction of thousands of hectares of forest every year.

Although this widespread view had not been empirically supported, it nonetheless had tremendous influence on forestry policies (Aguila, 1982). This was evident in the creation of regulations making forest occupants and shifting cultivators illegal. This perspective has, however, significantly changed since the 1970s. Several development programmes were initiated under the philosophy of transforming uplanders from 'agents of destruction to partners in forest development and conservation'. One such programme is the Upland Development Programme (UDP) of the Bureau of Forest Development (BFD).

There are two principal types of upland farmer: those belonging to tribal communities, and those lowlanders who have migrated into the mountains. These two types of upland community differ in environmental attitudes, behaviour and values regarding land ownership. Most tribal communities practice slash-and-burn cultivation, known as *kaingin*. They believe that customs and traditions based on historical patterns of land usage determine land ownership and boundaries. No one individual owns a particular piece of land and there is marked community cooperation in cultivating and maintaining the land.

By contrast, lowland migrants who have turned to shifting cultivation perceive the land they have claimed from the forest as their personal possession. Their greater experience with the market place has prompted them to change from the crops of lowland farming to those that can be grown in the mountains for cash. Compared to their tribal counterparts, these migrant settlers have a higher cash income, relatively new skills and farm implements, longer exposure to formal education and more material needs (Ganapin 1979).

73

Another contrast is that while tribal communities have intimate knowledge of natural phenomena and make restrained use of natural resources, migrants view farming as 'a production technology, (from which) to derive economic benefits' (Ganapin 1979).

The Upland Development Programme is working in three locations, involving both tribal and migrant farmers. The tribal community is in Luzon (northern Philippines) and we are also working with a Christian migrant community in the Visayas (central Philippines) and a mixed community of Muslims and Christians in Mindanao (southern Philippines). Two trained BFD personnel are based in each community, coordinating a programme on social forestry.

In the process of evolving models of upland development, the Upland Development Programme has conducted three kinds of activity in these areas: community appraisal, community organization and process documentation. The first of these concerns the collection of information specific to a pilot site for use in identification and planning of social forestry projects. Community organization depends on using the information so generated to organize groups of people within the community for the design and implementation of projects. Finally, the documentation of activities is a feedback process whereby problems encountered can be given immediate attention. In addition, documentation also provides a mechanism for drawing out lessons for upland development.

Community appraisal of the three sites described was conducted in 1984 and 1985 by a team of researchers from the University of the Philippines at Los Banos (UPLB), together with BFD personnel based at each site. The appraisal process was conducted in two phases: rapid appraisal and more comprehensive appraisal at a later stage. The community appraisal identified agroforestry as the main theme for projects in the pilot areas and the university team has provided a short training course on agroforestry for the BFD officers who are coordinating projects.

The university team is also continuing to work with farmers on problems in agricultural production identified during the community appraisal. Some farmers volunteered to try out specific improvements in technology and university scientists are involved in these on-farm trials. There is also a demonstration farm managed by BFD personnel with help from the scientists which shows contour farming techniques and tests different crop combinations.

Walking tours for rapid appraisal

At the very beginning of the work, four main rapid appraisal methods were used: walking tours for site reconnaissance, short social surveys, direct observation and group interviews.

A walking tour of the community is essential in the rapid appraisal of the upland situation and functions as an 'ocular survey' of variations in terrain, slope, production activities, vegetative cover and settlement pattern. The walking tour occupies the first visit to each site by the interdisciplinary appraisal team. In addition to the university researchers, the reconnaissance

74

team is composed of the project field coordinators (BFD personnel) and two or three farmers from different sections of the pilot site. Prior to the walking tour, a meeting of team members is held to discuss the aims of the exercise. Emphasis is placed on the identification of variations in physical, biological and social characteristics of the pilot site.

The farmers play a prominent role in the ocular survey. They lead the team to various sections of the pilot site pointing out its physical boundaries. They describe previous and current vegetative cover of the land, production activities and problems encountered by farmers. The researchers document the survey in pictures, ask questions and record the responses of the farmer members of the team. An output of the reconnaissance is a crude map of the pilot site prepared by the team. On a large piece of manila paper, the team indicates the relative location of different agroecological zones, political boundaries of the village, different social groups, community landmarks and infrastructure, sources of irrigation and potable water and access routes.

After the ocular survey of the site, a big community meeting is called. Almost all available farmers attend (together with their wives carrying infant children). In the meeting, the BFD personnel and the researchers explain the objectives and activities of the programme and the roles of the project field coordinators. There is also a long discussion on community problems, generally focused on land tenure and production systems. After initial hesitancy, the farmers can easily articulate their concerns and the wives contribute their opinions.

The significant issues, problems and concerns that come out of the meeting are noted. Project field coordinators and community leaders help identify key informants on these issues. These are persons who are knowledgeable on specific concerns by virtue of their role in the community, their length of residence in the community, or direct involvement in an issue. These informants are interviewed in depth for further elaboration and insights.

The rapid community appraisal used here includes a *social survey* because obtaining representative profiles of upland communities is one of the major concerns of the appraisal process. Thus, the interdisciplinary research team prepares for each site an interview schedule consisting of questions on socio-demographic characteristics, production and consumption activities, soil conservation techniques and social organization. The interview is pretested and translated into the local dialect of each pilot site.

For the selection of survey respondents, an area frame sampling design is used to provide a system for comparing information with respect to strata and to ensure that the sample is representative of the community population. Using data gathered in site reconnaissance, a stratification factor is identified for each of the three pilot sites. The factor for each is administrative and political boundaries, agro-ecological zones and segregation of social groupings. A complete list of farmers within each stratum is prepared and a simple random sample is drawn.

Respondents are interviewed in their homes. The interviewers are the university researchers and project field coordinators. In some cases, spouses assist the farmer-respondent in recalling the information sought in the survey. A few respondents volunteer additional information to that asked in the survey.

Direct observation of selected farms is complementary to the social survey. A smaller number of farmers are selected from among the survey respondents to represent each stratum, every cropping system and different soil conservation measures. Farms are visited by the researchers with farmers themselves serving as guides. Observational techniques are used for two purposes. One is to verify some of the data obtained from verbal responses of farmers to survey questions. The second purpose is to gather more information on land uses, agroforestry practices, cropping systems and measures used to control soil erosion. The farm visit is an occasion for further interaction between the farmer and researchers.

Small *groups of farmers* from each sampling stratum of the pilot site are interviewed. No selection procedure is used; the composition of a group depends upon the availability of farmers. They may or may not have been respondents in the social survey. Groups of three to six farmers are interviewed at a time. The group decides on the place for the interview. (In retrospect, we have noted that the houses volunteered as venues represent the relatively better built and bigger houses in the pilot site.)

Interviewing small groups of community members at a time has proved useful in discussing general subjects and community issues. These include use and availability of farm inputs, problems encountered in crop production and marketing, land tenure problems and social organization. A panel consisting of two researchers and one BFD official participate. The procedure provides the farmers with opportunities to check on and complement one another's responses. The interviewers, on the other hand, are able to follow up questions of colleagues from complementary perspectives.

The final stage is when the university researchers present the rapid community appraisal results and recommendations in a community meeting called for the purpose. The community members are given a chance to comment on the inferences drawn and on the interpretation of data. They participate very actively in the discussion of recommendations for development projects to be undertaken in the community.

Comprehensive appraisal

For the duration of one cropping season after the rapid appraisal has been completed, a series of further appraisal activities take place. The most important are periodical interviews of farmers who were respondents in the social survey and whose farms were appraised. The purpose is to ask questions and document their agricultural production activities, decision-making process, problems encountered, and ways of coping with problems. The comprehensive appraisal then leads to the construction of flow diagrams showing the production of major crops, with decision trees to analyse significant decisions.

76

In addition, researchers observe the farmers during land preparation, planting, weeding, harvesting and in some instances even the processing of crops, noting implements and inputs used. Construction of rock terraces to control erosion has also been noted. The folk rituals associated with crop protection and linked to local 'technical' knowledge have also been observed and documented, including, for example, procedures for the control of rats.

Comprehensive appraisal also includes use of a soil test kit, and under the supervision of the agriculturalist from the research team, the farmers themselves collect soil samples and do chemical analysis as well as measuring the depth of top soil. The outcome of the analysis is then compared with standard colour charts to determine the levels of nitrogen, phosphorus and potassium content of the soil. Farmers find the procedure easy, interesting and very useful in determining the type and amount of fertilizer to use on their land.

Other measurements taken by researchers working with farmers included the planting distance of major crops, the quantities of fertilizer and insecticide used and slopes and effective size of farms. This leads to much discussion of such things as the ideal planting distance for crops and contour farming.

Comprehensive appraisal leads on to project work in crop production and agroforestry. Here it is sufficient to sum up the appraisal activities as involving both direct and indirect interaction with farmers. In community appraisal, there are direct farmer-scientist interactions during site reconnaissance, social surveys, farm observation, group interviews, discussions of rapid appraisal results and then later in the periodic interviews associated with comprehensive appraisal and the soil tests. In the agroforestry work which has developed since, there are direct farmer-scientist interactions in on-farm trials and also indirect interactions through demonstration farms and via the BFD field officers.

2.5 Diagrams for farmers

GORDON R CONWAY

Diagrams for communication

A diagram or, expressed more fully, a diagrammatic model, is any simple, schematic device which presents information in a readily understandable visual form. Diagrams can radically simplify complex information, making it easier to communicate and analyse. Until recently, it has been widely assumed by professionals that rural people, especially when illiterate, would not be able to construct or understand diagrams. A mounting body of evidence, including that accumulated through agro-ecosystems analysis (Conway 1985; Conway et al 1987; Pretty, ed, 1988; McCracken et al

1988; McCracken 1988), indicates that not only can they often draw and understand diagrams, but that they take pleasure in doing so. We have found that diagrams can be an effective and efficient means whereby farmers' knowledge can be made explicit, either through diagrams they construct, or through their guiding and informing researchers so that they can make diagrams.

Diagramming in these ways has three major advantages over most other modes of investigation:

• the questioning and responses are more open-ended than in, for example, questionnaire surveys. In diagramming, the general subject area may be preset, but the detail has to be filled in by respondents, giving primacy to their knowledge and perceptions;
• diagrams can capture and present information which would be less precise, less clear, and much less succinct if expressed in words. This makes practical analysis easier;
• diagrams are shared information which can be checked, discussed and amended. If diagrams are drawn up by researchers during interviews, respondents can examine what has been recorde on the basis of what they have said, and confirm or qualify it.

Our emphasis is on the use of diagrams derived mainly from local knowledge as a tool for communication and analysis for agricultural research. Five types of diagrams will be described.

Maps are an obvious and simple type of diagram. Their primary use by farmers is in communicating with development specialists, particularly about the location of different parts of a farm, their relationship to basic resources, such as water, and to the major land forms.

To most westerners, maps are readily comprehensible but may be foreign to some cultures. In that case, techniques of map construction need to be tailored to local perceptions. The best approach may be to ask people to draw their village and then build on the conventions for representing its layout which they adopt. Obviously, the western convention of always having north at the top of the map may be a hindrance to understanding in some circumstances. Joint map making with researchers and farmers needs to begin with one or two commonly agreed reference points, the remainder of the map being constructed on these. In Pakistan our practice has been to construct sketch maps from a high vantage point, using this approach. Other uses of mapping methods by farmers are discussed in section 2.6.

Transects (see figure 2.1) have greater practical utility than maps, in our experience. They can focus attention on the different zones or micro-environments in a watershed, village or farm. In agroecosystems analysis, they are drawn up by researchers who walk from the highest to the lowest point in an environment, accompanied by local people, consulting people in each zone. The main purpose of transects is to identify the major problems and opportunities in the agroecosystem, and where they are located. Transects need to be simple, indicating the major topographical

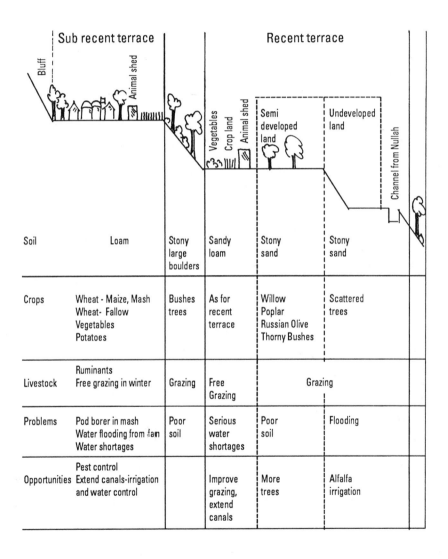

Soil							

Figure 2.1: *Transect of a village in northern Pakistan*

79

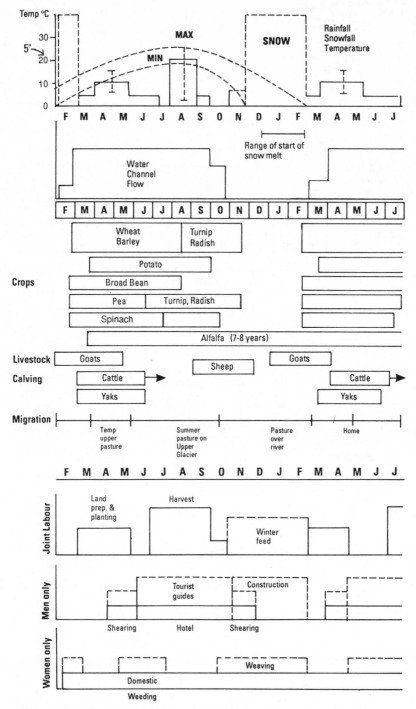

Figure 2.2: *Seasonal calendar for a village in northern Pakistan*

80

features, with associated lists of the crops, livestock, problems and opportunities (as in figure 2.1). If soils are included they should be referred to by some feature such as texture or water holding capacity. As with the maps they do not need to be cartographically accurate. Each transect should be a schematic idealization, not even necessarily following a real straight line. Although they have some similarity with botanical or soil survey transects they are different in form and purpose.

Calendars (see figure 2.2) are diagrams to indicate seasonal features and changes. Where possible, they should be based on the local calendar. They are useful to enable farmers to identify critical times in, for example, the annual crop cycle. Calendars have long been used in Farming Systems Research. They can be used to cover all the major events and changes that occur within the rural year. The most obvious and important dimensions are climate, cropping patterns, livestock (sources of forage and key events such as calving, sales and migration), labour demand, diet and nutrition, diseases and prices for crops, livestock and other produce and for food.

Climatic data may often be available from official records, but farmers' own perceptions can be valid as well as indicating the view of conditions on the basis of which they make farm decisions. In northern Pakistan, where rainfall data is largely absent, we found that semi-structured interviews can give relative rainfalls. Questioning goes approximately as follows: Which is the wettest month? Which is the next wettest? and so on, followed by: Which is the driest? Which is the next driest?, together with comparisons of months: Which is the wetter (or drier) of these two? Relative amounts can be roughly gauged by asking comparisons of wetness, whether one month is three-quarters, a half or a quarter as wet as the wettest month. Relative amounts are adequate for initial diagnostic purposes, showing the pattern into which crops have to fit. It may be enough simply to construct a seasonal calendar around major events such as the beginning of the rains, periods of drought, first frosts or the level of irrigation canal flow.

Rural people's knowledge of climatic events can be detailed. In an interview in Wollo in Northern Ethiopia (ERC 1988), two farmers recalled the number of days of rainfall in each month for the previous five years. Their recall probably picked up rainfall which was agriculturally significant and may therefore have had an agricultural validity superior to that of normal rainfall records. In any case, their achievement shows the value of assuming that rural people have detailed knowledge and asking them about it.

Agricultural labour demand for women and for men can be elicited in a similar manner, asking first about the busiest and the next busiest months and so on, then the least busy, and so on. In West Bengal, the resulting histograms for women and men have been drawn on the ground, provoking a debate about the different labour peaks for women and men and the continuous labour demand, pointed out by the women, of their domestic chores.

Visually, 18 month calendars are better than 12 month calendars for

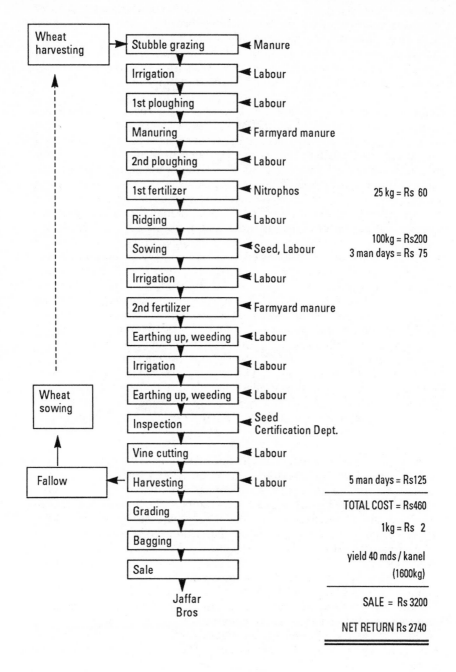

Figure 2.3: *Production cycle for seed potatoes*

82

revealing seasonal patterns. Conventional Western calendars begin in January, but the local calendar may start at some other time, or it may be better to start with a key event like the start of the rains.

Flow diagrams (see figure 2.3) can elicit and present sequences such as the cycle of production and marketing. Key aspects can be noted alongside the flow, for example labour requirements and monetary costs. These then become simple production accounts.

Venn Diagrams can also be used, for example for understanding institutional relationships within a village. Even in quite small villages the number of different institutions and actions involved in decision-making can be considerable. These can be identified and diagrammed at a meeting of villagers or of a particular group. Venn diagrams, known in Pakistan as 'chapati diagrams', use touching or overlapping circles of various sizes. Each circle represents an individual or institution and the size of the circle indicates importance (which can be discussed by the group undertaking the exercise). The circles can be used to indicate the degree of contact or overlap in terms of arriving at decisions. Overlap occurs if one institution or individual asks another to do something, or if they have to cooperate in some way, according to the following convention:

- separate circles mean no contact;
- touching circles indicate that information is exchanged;
- small overlaps point to some cooperation in decisions;
- large overlaps mean considerable cooperation.

Diagrams to aid analysis

Beyond their use to elicit information, diagrams can be used by and with rural people as an aid to analysis (see also pp 93–100). Most of those described are general purpose tools for identifying problems, constraints, solutions and opportunities. Most of them have been used in group discussions by teams of researchers or extension specialists or combinations of these (Conway 1986) but have recently been extended more to aid analysis by rural people themselves, as the following examples illustrate.

Seasonal diagramming can focus attention on key seasonal constraints. In one village in Pakistan, for example, systematic seasonal diagramming revealed that the period when dysentery was rife was also the time of harvesting, posing a problem with an agricultural as well as a human aspect. Similarly, seasonal diagramming in South Wollo in Ethiopia found that the peak month for malaria was also the month of highest male labour requirement, for land preparation. More positively, the analysis of seasonal diagrams can point to opportunities, such as when new crops can be grown.

Venn diagrams for village institutions can similarly be used to generate and focus analysis. We have found that they can be constructed very easily by cutting out paper circles of different sizes, labelling them with the names

83

of the institution or individual and then arranging them on a table in a pattern that emerges from the discussion and experience of the participants. Once arranged to everyone's satisfaction, the circles can be stapled into position and used to identify needs for improved links, better overlap, or the positioning of new institutions.

The power and utility of diagramming can, finally, be illustrated from a workshop carried out in the Philippines which focused on a small dam at the outlet of Lake Buhi in Bicol province (Conway and Sajise, 1986). Following construction of the dam a number of severe problems arose, primarily affecting the lakeside dwellers and inhabitants of the municipality of Buhi. Those adversely affected became understandably angry, to the point that the future of the project was in jeopardy.

In order to try and tackle the problem a team from the University of the Philippines undertook a brief survey of the area, interviewing farmers and summarizing these interviews, together with secondary data and direct observations, in a series of diagrams similar to those discussed above. A four-day workshop was then convened which brought together some 70 people representing not only the aid agencies and the central and provincial government agencies, but also local politicians and representative farmers and fishermen. The workshop was aimed at conflict resolution, using a procedure for the analysis of diagrams. This worked extremely well and one of the most satisfying memories of the workshop was of intensive but productive arguments between small groups of aid and government officials and farmers and fishermen, focused on a particular diagram (figure 2.4).

It turned out that one of the key diagrams was a seasonal calendar which helped resolve the central water scheduling issue. Fishermen above the dam were complaining of their fish cages drying out and lakeside farmers of their rice fields suffering drought, in order to provide water for the downstream farmers. Construction of the seasonal calendar pinpointed the key constraints to the timing of agricultural and fishery operations, namely the occurrence of typhoons and sulphur upwellings, but also demonstrated that retaining the water in the lake above a critical level until the end of May could satisfy the upstream farmers and fishermen without severely affecting those downstream.

Many of the examples cited earlier, however, refer to work being undertaken by the Aga Khan Rural Support Programme (AKRSP) in northern Pakistan (Conway et al, 1987). This programme aims at rapid development in several hundred villages primarily through the efforts of the villagers themselves. However, it remains to be seen how far diagrams will be used by village organizations. One suggestion is that valley-wide groups may use agro-ecosystem zoning for planning, and the mapping of agro-ecosystems already done has been very illuminating.

Traditional land-use capability classification and agro-ecological zoning tend to be data-hungry, requiring extensive field surveys and information on climate, soils, vegetation, etc. By contrast, the method used in northern Pakistan is meant to be rapid and iterative. In 1987, a trial zoning exercise was under way in the Hunza valley, with a first rapid survey

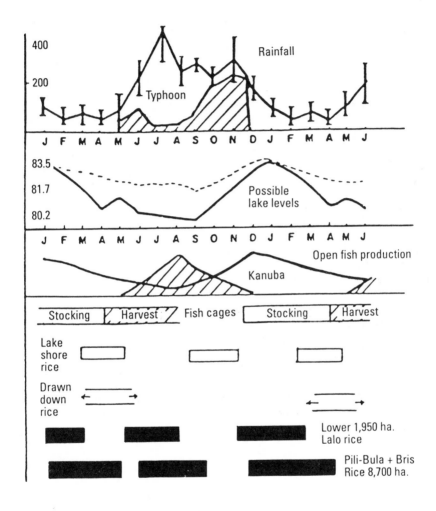

Figure 2.4: *Seasonal calendar for the Lake Buhi project*

85

covering biophysical features. An iteration emphasizing socio-economic features was to follow and it was assumed that boundaries would be revised as more information was gathered in subsequent iterations. Biophysical zones were characterized by the growing period for crops, which varies with altitude from 190 to 330 days. Calculations were made from secondary data, but then interviews with farmers were used to alter and refine the initial boundaries. Farmers were well aware of the major differences we were mapping along the valley and could give us accurate estimates of growing periods in their own and neighbouring villages.

A second technique under development is sustainability analysis. This is a group exercise which can be used to investigate either a particular production and marketing flow or a development process or project. The aim is to identify problems and threats that are likely to arise and to think of solutions. The exercise begins by construction of a flow diagram. The groups sit round a very large sheet of paper on which the flow diagram is marked out in black. Then participants use coloured pens to mark in problems that they know or guess will arise (in blue), stresses or shocks that may occur (in red) and suggested preventive measures or solutions (in green). Finally, points in the production cycle are marked in yellow where checks (or monitoring) seem advisable to see if problems are developing. It is hoped that village groups will find it useful to analyse sustainability of key production processes this way.

Conclusion

The potential of diagrams for eliciting the knowledge of rural people and for analysis by and with them, is only just beginning to be realized. Professionals concerned with rural development have tended to suppose that rural people, especially if they are illiterate, will not be able to understand or use diagrams. Our own experience has been that their capabilities almost always exceed the expectations of outsiders. The best rule of thumb seems to be to assume that they can understand and use diagrams until proved otherwise.

For the future, there is scope for much inventiveness, by rural people and by outsiders, in devising and using these and other diagrams, exploiting the advantages they have over more conventional methods of investigation and analysis.

2.6   Maps drawn by farmers and extensionists

ANIL K. GUPTA AND IDS WORKSHOP

Different views of reality

Mapping can be viewed as one specific type of diagramming method, as noted by Gordon Conway in section 2.5. It might appear surprising to include maps as part of this discussion of innovative methods, since mapping is a conventional technique which has been used for decades as a part of formal agronomic and geographical research, land-use planning, and a guide to many other activities, and it is undoubtedly a vital tool for many purposes. However, while recognizing the importance of mapping in its conventional forms, our discussion instead stresses new approaches.

Reality mapping is a method we have used in several contexts, most recently in India[1] in July 1987, as an attempt to understand the way poor people perceive their environment. What we do is give coloured pens and paper to individual men or women on an occasion when they are meeting as a group, for example, in a women's workshop. Very often, these are people who have never handled a pen before. We then ask them to draw their village or any aspect of it which they see as important to their survival. Not infrequently, at least among women, the result is only a colourful pattern or design without any recognizable figure or shape. However, many do draw trees and plants and, almost without fail, a temple. It is instructive to study which species they draw and what plants they feel most comfortable about drawing.

We have tried this exercise with Indian Administrative Service officers in India and scientists in Bangladesh as well as with poor women and farmers. Several differences reflecting social background and sex have emerged. Women tend to draw very small forms, often centred on a temple and rarely showing any means of transport. Men, by contrast, rarely omit transport. Moreover, in the case of a dry village in Maharashtra where a student named Mandavkar tried it, it was noticeable that while many poor people drew only their immediate neighbours and fields, the richer people drew in far more detail covering the whole village.

We are still analysing what may be gained in understanding people's cognitive maps through drawing, and we do not claim much at present. However, we can recommend its use as an 'unfreezing device' at the beginning of farmers' or scientists' workshops. It can generate tremendous creativity in group sessions, as happened at a session at Bangladesh Agricultural Research Institute (BARI). There paintings done by groups were bold and satirical, in contrast to the more restrained individual ones. What we hope to do, however, is to refine the technique so as to help us understand how people relate to their resource environment.

Topographical maps by pastoralists

In some places, village people have been encouraged to draw maps of a more strictly topographical kind. One agency in Mexico regularly asks people to draw maps of their own villages. An instance in West Africa relates to Jeremy Swift's work with Wodaabe Fulani pastoralists in Niger (Chambers 1983:99). The research team asked some herders to draw maps, which they did without difficulty. The maps differentiated areas according to their ecological characteristics, as might have been expected, but also indicated several special zones. These were areas where the herders' cattle got night blindness from vitamin A deficiency in the dry season, for which reason otherwise good pastures had to be left unused. The herders associated night blindness in these areas with the absence of green plants which scientists could identify as sources of carotene (from which vitamin A is synthesized in the body). It emerged that staff of the livestock service, which had been operating in the area for 50 years, were totally unaware of the problem. Thus a significant item of veterinary information, well known to pastoralists, was brought to the attention of official science by a simple mapping exercise.

Subsequent experience was also revealing. The research team obtained some vitamin A and took it to a Wodaabe camp where the cattle had night blindness. A cattle owner there was willing to let his animals be treated, but asked for only half of them to be given the vitamin A so that he could observe the effects and compare them with the untreated half.

Another means by which farmers or pastoralists may draw maps is by working in a group led by a researcher or extensionist. The latter may do the actual drawing, but following instructions shouted out by the group. In Bangladesh, one of us asked farmers to tell us of areas with particular problems so that they could be marked on a map. In north-east Kenya, working with a group of Somali pastoralists, another of us mapped water holes by asking about distances from a number of known fixed points and already marked on official maps. Distances were agreed by the group in terms of the time it took to walk and once the distances of a water-hole from three of the fixed points had been established, its approximate position could be marked in on the map. Then soils and vegetation were also marked and a map was produced which could help in understanding and moderating conflicts between camel and cattle owners.

It is probably significant that two of these examples concern pastoralists who are always travelling. Such people may tend to perceive their reality in terms of the surface of the earth, and may thus adjust to the conventions of maps rather easily. The Indian women who drew temples but never means of transport would possibly have mental maps less adaptable to flat sheets of paper, and perhaps they always stay so close to home that they would lack data to put on topographical maps.

Other mapping concepts

Arable farmers may think about the areas of land they use in yet another way, not being extensive travellers like pastoralists, nor yet as localized as many village women. Maurya and Bottrall (1987:19) imply a good deal about this, although they do not cite any drawn maps, when they suggest that farmers in India categorize their fields according to a topographical system with which they associate rice varieties. Farmers have 'segmented' thousands of indigenous rice varieties into a few classes related to different agro-ecological situations, so that they say, for example, that *Bhadai* rice grows on sandy upland, *Kwari* on upland loam, *Kartiki* on medium land, and *Aghani* on water-logged ground and in deep-water situations.

It should never be assumed without careful testing that non-literate people cannot understand or use maps. Drawing maps in conjunction with farmers, whether literate or not, can, to the contrary, be a way of enabling them to share their local knowledge. Scientists can easily fail to identify agroecological zones through not consulting local people. Sometimes an area which seems homogeneous contains micro-environments crucial to local farming systems. Examples include homestead gardens (see index) and riverine strips and dambos in Zambia (pp 32, 125), and even a distant tree may be part of the 'property' of a homestead, playing an important role as a source of fruit, fodder, or other material. Such significant areas and resources, though small, are more likely to be picked up in local people's maps than in those of outsiders. Besides ecological factors, maps can be used to indicate access and control: who, within a household, owns a particular field, and who in the community manages particular areas of forest or grazing.

Aerial photographs may usually be more readily understood (unless they were taken at peculiar angles, or in false colour, or are out of date) than maps and can be a useful aid to dialogue. Indeed, many farmers are fascinated by aerial photography and one of the nicest gifts one can take to a farmer is a photograph which shows his or her land. In Costa Rica, it has even been found possible to use satellite photographs to show villagers how the environment has changed over the past 20 years. Reduction in forest cover, for example, can be vividly demonstrated in this way.

Two cautionary points must be noted, however. When villagers are first shown aerial photographs, some explanation will be needed as to their context and scale. More important, since the easiest way of orientating the photographs is by reference to roads, any changes in the road system since an old photograph was taken may cause difficulty.

The idea of farmers and researchers working together on a map can be applied to other forms of diagram, of which several examples are discussed by Gordon Conway in section 2.5. Perhaps the most generally useful is the kind of diagram which is sometimes loosely referred to as a 'systems map'. This can exist in many forms, such as the simple representation of 'the components of a typical small farm' used by Charoenwatana (1987) or the 'chapati diagram' illustrating institutional relationships in a village suggested by Conway. Such 'maps' of social and ecological relationships tend to start by seeming simplistic and end by being over-complicated. What is important, however, is the process people go through in making them, and in figure 2.6, Clive Lightfoot and his colleagues show a systems diagram which was the outcome of a process in which researchers and farmers worked together, sharing knowledge on means of tackling a persistent weed.

Environmental mapping[2]

When geographers make maps to analyse the physical characteristics of a region, the emphasis is on objectivity and cartographic accuracy. When making maps with farmers or extension workers, our starting point must be that a watershed, village or district may be seen quite differently by different people. Recording the contrasting perceptions of researchers and extension workers can expose their different assumptions about farmers' problems. Such contrasts are brought out when impressionistic rather than cartographically accurate maps are prepared to show the ecological niches occupied by different crops, technologies, enterprises or management practices. Making maps can thus provide a medium for exchanging ideas about the respective ignorance and knowledge of each group involved.

Extension workers who have been travelling and interacting with farmers in a region develop extensive insights concerning local ecological endowments and limitations. Village development workers employed by the extension department (eg, in Bangladesh) may also have a very precise understanding of the interplay between edaphic and climatic choices confronting farmers. In making these choices, farmers have a great sense of space and season and their relationship to economic sectors. Over time, however, many of their decision-making rules may have become rituals, so that avowed reasons for a choice may not correspond with the 'real' reason. These things have to be carefully distinguished, but our view is that physical and ecological factors determine the niches within which a particular technology is used much more strongly than the socio-psychological factors on which so many theses have been written.

The first step in getting extension workers to prepare ecological maps is to obtain outline maps for the district or village or 'village-level-worker' area concerned. Several copies are needed by each extension worker and researcher taking part. The maps are marked only with geographical boundaries and no other detail.

Next, each participant is given a set of maps and individually fills them up, showing the extent of different crop varieties and farm enterprises, using agreed symbols to depict the maximum and minimum extent of each. The maps are drawn by juniors as well as seniors, extensionists and researchers alike. The purpose of working individually at this stage is for each person to become aware of his or her respective strengths and weaknesses. Each map thus shows the area where the maximum proportion of the land is under a particular crop, or has other uses, and where possible, dominant different crop varieties are also shown. Maps for livestock, trees and vegetables as well as the main crops can also be prepared. Risk maps showing frequency of drought or flood can also be prepared separately.

Once these maps have been prepared, there are two options. One may either have a workshop to consolidate the individual maps by negotiating about the veracity of respective claims, or else use small groups to consolidate maps around different kinds of perception. One reason for differences between different people's maps is that many boundaries are not sharp and indeed may change from year to year, following variations in weather. Shifts in areas planted to a particular crop may be quite substantial. However, the range of these shifts can be marked on the maps. For example, in Bangladesh, after floods in the *Aman* paddy season, the area under the succeeding wheat crop generally increases because of good residual moisture and early availability of the fields due to premature harvesting of the *Aman* rice.

When consensus maps have been evolved in this way, they have proved to be reasonably precise. The professionals further down the line have been found to be usually more exact in their mapping than those at higher levels.

The finished maps are next discussed by a panel of farmers, who are asked about the niches occupied by different varieties, crops or tree species. The insights offered are sometimes so precise that one marvels at the farmers' sense of space and season. They are able to say why certain enterprise combinations are localized only in certain niches and under what conditions these niches may change. Since farmers do have such clear understanding of their environment, we strongly question the assumption that what they need to be told by extensionists is just practical information about technology, without reference to the science underlying such advice.

Farmers' comments sometimes reveal that enterprises may be eco-specific in one context and class-specific in another. For instance, the sweet potato is cultivated by only the poorest people on the uplands in Bangladesh, but on riverine lands this crop would be found suitable by everybody regardless of class. Likewise, mixed cropping is generally practised by the poorer households, but on uplands and in regions growing sugar cane, it is adopted by most people.

Another instance of class specificity can be seen in the variations in the use of an enterprise rather than its extent. The poor may use sweet potato as staple food during stress periods while the rich never would. Skills involved in an enterprise may not just be class specific, but also caste and gender specific. The derooting of sweet potato vines is done only by women, who consider it helpful in producing round tubers which fetch better prices than the usual long ones. Likewise, sisal is cultivated on the bunds of fields and on poor soils in semi-arid western Maharashtra and the skill for processing it into fibre is restricted to the Mang lower caste group and the landless community.

Ecological mapping based on local knowledge thus helps in delineating the ecological and management context of different technologies. Anil Gupta has found that ecological mapping is a good diagnostic tool for understanding the relationships between space, season, and sector (such as crops, livestock, and tree species and varieties). It has practical applications especially in rainfed regions where technological breakthroughs have been elusive. Mapping the areas under rainfed crop varieties provides insights about the boundaries of agroecological niches. These then help in targeting trials of prospective technologies to those niches which are most suitable, in contrast with the common practice of locating on-farm trials on land that is convenient for access. This approach, he argues, is more effective and convenient than those of extrapolation area and recommendation domain developed by IRRI (Zandstra et al, 1981) and CIMMYT (Byerlee and Collinson, 1980). Ecological mapping can also suggest reasons for the limited diffusion of proven technology, and can help towards strategies for the promotion of new crops, varieties and techniques.

Some of the most significant information to come out of ecological mapping in Bangladesh concerns the relationship of risk, environment and poverty. Mapping famine vulnerability has been attempted by geographers. Generally, too, the poorer households are located in the most vulnerable locations, with the greatest number of the poor living on river banks and riverine lands. Until ecological mapping was attempted, this association of high risk and poverty was not always recognized, for example, in vulnerable regions in Ishurdi.

In these ways then, ecological mapping can help in setting priorities for research, with respect to problems tackled (in terms of enterprise, sector, season or space, ie, niche), with respect to how and where trials should be conducted and with respect to risk, vulnerability and poverty.

Mapping for agronomic monitoring

Mapping can also be used for agronomic monitoring. As described by Richard Edwards (1987c), the aim of agronomic monitoring is to provide a 'picture of what is happening on farmers' fields' and to relate this to the resulting harvest for each crop. Information obtained by observation and by asking questions of the farmers is plotted on a map by the researcher 'in a rational and systematic way'. Edwards gives an interesting example of the potential value of this procedure. In one part of Lusaka Province in Zambia, variations in residual and seepage moisture in the soil, coupled with very warm ambient air temperature, mean that maize is planted sequentially on different sites throughout the season. 'By mapping the farmers' operations, it was possible to study the build-up of various diseases, in particular maize streak virus. With this knowledge, it is possible to say to the maize breeders that there is a particular 'hot spot' which can be of use both for testing material and for the collection of germplasm, as there is the possibility that resistant material may be common in the area'.

Mapping for agronomic monitoring requires that fields are observed regularly throughout the season, and the dates of various events plotted. In this case, the dates of maize planting and of the emergence of symptoms of maize streak virus would be most relevant, but in other examples Edwards quotes, information about land preparation techniques used by farmers emerged from plotting the relevant operations prior to planting.

In discussing other uses of agronomic monitoring and their values, Edwards notes that mapping seems to be most effective when undertaken by all members of a multidisciplinary team, including agronomists and social scientists. One advantage is that it helps the team to work well together.

2.7 Systems diagrams to help farmers decide in on-farm research

CLIVE LIGHTFOOT, OLIMPIO DE GUIA JR., ANICETO ALIMAN, FRANCISO OCADO

Farmer–researcher interaction

In our recent research project in Eastern Visayas, The Philippines, systems diagramming was used as one of several effective methods to facilitate researcher–farmer discussions, to identify farmers' practices and priority problems and, consequently, to help towards developing on-farm experiments. Activities comprising the method – farmer–researcher interaction, systems diagramming and screening solutions – are discussed briefly below.

Both group meetings and individual farm visits were used to facilitate three rounds of discussion between farmers and researchers, which allowed the farmers' priority problems to emerge. In the first round, various topics were discussed in group meetings, during which farmers selected issues they wished to elaborate further. At first they discussed issues of credit and seed supply, probing to find out whether the researchers would provide free inputs as we had in the past. Once they realized that nothing was going to be doled out, declining soil productivity was emphasized (in animated conversations) as a major concern. Researchers listened as farmers explained their current land-use systems and cultivation processes, and heard for the first time about shifting cultivation in the area, fallow rotations and classifications related to this system.

Enough interest was generated by these discussions for some farmers to ask us to visit their fields and in that way they led us into round two. This consisted of separate one or two hour discussions with four individuals on their farms. In these meetings, we began to appreciate the complex web of issues involved in the seemingly straightforward subject of soil productivity. During the farm visits, we saw that most areas were dominated by cogon grass (*Imperata cylindrica*). Even though farmers knew this land was poor they were forced to cultivate it because better areas were too far away and population pressures were increasing, or because it was the only place suitable for subsistence crops.

Round three brought everybody together again for a group meeting to obtain consensus on a priority problem. (A key point for a meeting at this stage is that if consensus is not attained, interest and cooperation will soon wane. Waning interest is, however, a useful check for research error.) From the complex of issues mentioned by our farmers in the last two rounds, consensus began to form around issues pertaining to the cultivation of cogon areas. Issues included cogon control, declining soil fertility and high labour and draught costs. The group wanted us to help them solve their cogon problem. However, before beginning research we needed to ascertain how relevant this problem was to neighbouring communities. This we did by a diagnostic survey covering 24 randomly selected

93

households out of 150 in three parts of Gandara Municipality, Samar. The survey was conducted on a basis of 'guide topics' (Table 2.1), which were identified by researchers and key farmer informants, referring to information gained from previous conversations with farmers about their use of the cogonal areas. In this survey, the interaction between farmer and researcher was more like a conversation than a formal interview. To maintain free-flowing conversation, topics were neither discussed in order nor finished in one session. This approach led us to visit specific parts of the farmers' lands which they had invariably requested to enable them to explain what was going on. All this added up to a lot of time and needed a lot of patience.

In that time, however, we listened and learned a great deal. Not only were farmers deriving income or making savings from on-farm activity, but also from a wide range of off-farm activities. Half the farmers earned money from ploughing and/or harrowing other people's fields. Of the farmers we listened to, only 13 per cent did not have such activities. Furthermore, their own farm enterprises included a broad range of livestock – carabao (buffalo) (68 per cent), pigs (50 per cent) and chickens (30 per cent) – and they cultivated several parcels of land.

Table 2.1: Guide topics used in the informal survey of 24 farmers in Gandara, Samar, the Philippines.

Farm typology
Household size
Farm size
Number and size of parcels cultivated/fallowed
Tenurial status for each parcel
Crops grown
Livestock raised
Sources of off-farm income
Cultivation practices
Criteria used for selecting cogonal areas for cultivation
Procedures for opening cogonal land
Sequence of crop planting in cogonal areas
Main resources used
Social structures for labour use
Rationale for cultivating cogonal areas
Reasons for cultivating cogonal areas
Reasons for using those cultivation procedures
Reasons for using those planting sequences
Factors that limit cultivation
Reasons for cogon being in the area

We also learned that their crop enterprises were very complex. Indeed, at some time or another our respondents cultivated up to four distinct agro-ecological zones: the sloping forested areas, rolling cogonal areas, flat upland areas and bunded areas used for rice (see figure 2.5). But for more than 60 per cent of them, individual parcels of land were less than half a hectare. Scattered across four agro-ecological zones, these small cultivated parcels were used under an assortment of tenure arrangements. Most forested land (93 per cent) was 'in the family' or rented. The poorest land, covered in cogon, was mostly owner-operated (35 per cent) or owned within the family (26 per cent). For more than half our respondents, fallow areas were dominated by cogon, with shrub species and other grasses. They brought such land back into production using both hand cutlass (*bolo*) and plough cultivation. Whichever way they do it, two or three months of labour costing around 1,500 pesos for one hectare is an enormous outlay for a person whose annual income is around 4,000 pesos. Why farmers continue to make these outlays but yet the cogon continues to thrive is something farmers explained through a systems diagram.

Systems diagramming

The systems diagram made a picture of the issues and interactions which farmers perceive as involved in the cogon problem. Our informal survey responses provided information on the biophysical causes and socio-economic constraints associated with the problem. Each cause or constraint was ascribed a box on the blackboard with arrows leading to the centrally placed problem box (figure 2.6).

Five key informants then met with us and explained relationships between boxes, often adding more, and indicating the relationship with the central problem. Then the boxes were redrawn into concentric rings around the problem, with each box forming one segment of a circular systems diagram. The size of each segment was determined by the proportion of farmers who responded to that point in the survey (figure 2.7). Finally, a group meeting of all respondents was called to obtain agreement that this systems diagram represented what happened on their farms.

Farmers told us that cogon was around mainly because seed blew in from surrounding fallow areas and easily germinated on the exposed infertile soil. Seed, soil exposure and fertility are thus represented as the three right-hand segments of the inner ring in the diagram (figure 2.7). The outer ring represents underlying causes of these conditions, which the farmers explained by saying that the soil was infertile because too many crops had been taken off under almost continuous cropping and a lot of erosion occurred when it rained. The soil was exposed because the area was always being accidentally burned, there were no shade plants around and intensive tillage practices kept the area bare. Such were the biophysical causes advanced by our respondents. On the other side of our systems diagram farmers sketched their socio-economic constraints to solving this problem. A lack of labour, draught power and land constrained them from

95

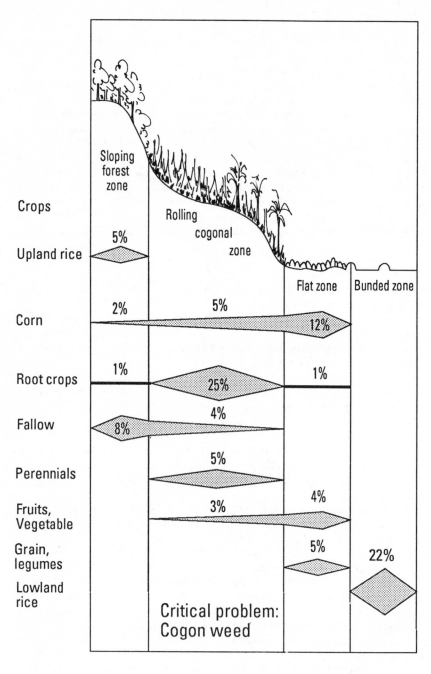

Figure 2.5: *Transect of land in the area around Gandara, Philippines*

96

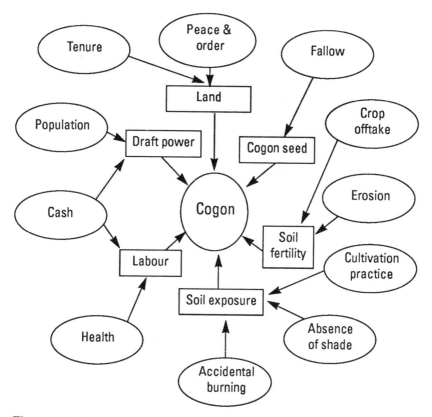

Figure 2.6: *Causal relationships drawn by farmers for cogon problem*

controlling cogon. Underlying the shortage of labour was a mixture of poor health in the family and no available cash to hire labourers. Cash also affected their ability to hire draught animals. Although land appeared under-utilized, access to it was restricted by tenure, there being no cash to pay rents. Peace and order and population pressure as farmers moved nearer the village for added security only aggravated the land shortage. With limited land, labour and capital, what kind of solutions could farmers test to rid them of cogon?

Screening potential solutions

Farmers used their systems diagram in figure 2.6 to identify possible places for solving the problem. In our continuing effort to let the farmers' priorities and ideas come first, we returned to the farm to elicit what experiments, ideas, or knowledge they had to offer. That farmers conduct experiments had already been determined by our site team in Gandara (Lightfoot, 1987). In these five farm visits, apart from discussions, researchers were shown interesting 'experiments' and natural phenomena.

97

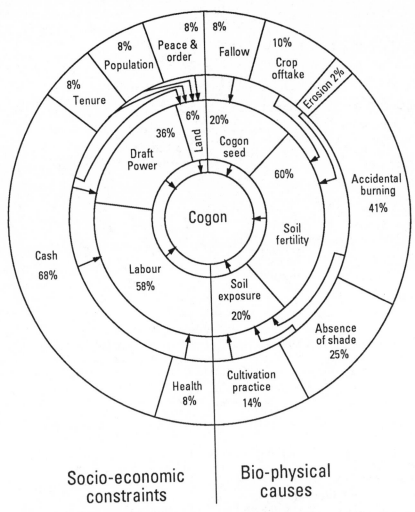

Figure 2.7: *Systems Diagram of the Percentage Distribution of Socio-Economic Constraints and Bio-Physical Causes for the Cogon Problem of Twenty Households in Gandara, Samar*

Several of our key informants had observed that cogon was shaded out or suffocated by vigorously vining plants like Kudzu (*Pueraria phaseoloides*) and Kurumput (*Passiflora foetida*). Most farmers knew that cogon neither grew in shaded areas nor germinated in shaded or covered soil. These observations were supported by formal research findings. Mercado discusses the importance of shade in the control of cogon (Mercado, 1986:268–278). Similar evidence was found by Sajise in his work on plant succession in cogonal areas of the Philippines (Sajise, 1984:141–153). Farmers also had other ideas for controlling cogon. Ploughing and planting cassava or sugarcane were two examples. Supplementing this list researchers advanced the idea of using herbicides. All these ideas were presented to the group for them to screen what they would like to test.

Key informants and researchers presented their various options whereupon the pros and cons were openly debated. The systems diagram was used to focus the debate; 'pros' became potential benefits vis-á-vis biophysical causes and 'cons' became potential conflicts vis-á-vis socio-economic constraints. For example in our meeting, ploughing and herbicides were deemed inappropriate solutions in view of the noted socio-economic constraints.

Money and labour constraints did not, however, appear to conflict with a potential solution in shading out cogon by planting trees or vining plants. Although several farmers wanted to try shading others wanted to see some of these plants growing before they would make any decisions on a field trial. A field trip was arranged for them to see Ipil-Ipil (*Leucaena leucocephala*) trees growing in contour hedgerows for erosion control and Kudzu (*Pueraria phaseoloides*) growing as a cover crop under coconut on some farms. One key point here is that if strong consensus cannot be reached and if all farmers do not rally behind any well defined test then it is better to go back a few stages rather than push forward an unpopular experiment.

After the field trip support formed behind the testing of vining legumes to shade Cogon. This support was further boosted when farmers traced out its system interactions. Specifically, they had a hunch that legumes may also directly improve soil fertility. Moreover, we also indicated that other legumes were known to nodulate more than Kudzu and thus might provide even more fertility to the soil. Formal research had found that *Pueraria* fallows increased soil organic matter and yields of subsequent corn crops compared to grass fallows (Jaiyebo and Moore, 1964). Furthermore, farmers thought that ground covered by a flat bed of legume would be less laborious to cultivate than that of tall grasses and shrubs. Thus, farmers were working on the hunches or hypotheses that a sequence of legume species will first control cogon and then regenerate soil fertility and also be easier to cultivate.

The biological basis for their hypotheses was that *Pueraria* and *Centrosema* species established by broadcasting seed over underbrushed or burned cogon would smother and control cogon. The resultant legume dominated vegetation would then be underbrushed and seeded with *Desmodium ovalifolium* which with its higher nodulation activity would relatively quickly regenerate soil fertility. In addition, this legume dominated sward should require less labour than that needed to recultivate dense stands of cogon.

Today, seven months after these discussions, 31 farmers have now identified experimental areas, 25 have established the first legumes – *Pueraria* and *Centrosema* – and 20 are currently starting nurseries for the other legume species – *Desmodium ovalifolium*. Even though legume growth was stunted by a long drought and legume cover was still less than 25 per cent farmers from neighbouring communities have

99

requested us to help them find planting material to start their own experiments.

Conclusions

Our contribution to the development of this farmer-first paradigm has been to provide a method for letting farmers decide research topics. This strategy is the missing element that Chambers and Jiggins noticed in 'Coming Full Circle' (Matlon et al, 1984): 'Only one element seems missing: a decisive and categorical specification of a process which encourages and enables resource-poor farmers to indicate what they need and want.' (Chambers and Jiggins, 1986:15). Our strategy – problem identification, systems diagnosis and hypothesis elaboration – minimizes the biases of our own agenda and maximizes the concerns of farmers. At the heart is the systems diagram. The diagram requires researchers to listen to farmers; indeed the diagram could not be drawn without the farmers' inputs. The diagramming exercise forces the union of diagnosis and testing, which are so often dislocated.

Gandara farmers are now solving a problem which could not have been solved by component 'transfer of technology' procedures. The use of the systems diagram has helped us to break out of the mental prison of commodity component thinking, and has led to 'unthinkable' research into such topics as enriched fallows and live mulches (Repulda et al, 1987; section 3.4). As a result, an appropriate agenda for research has been formulated, with practical action for farmers, researchers and extensionists alike.

2.8   Interactive research

IDS WORKSHOP[3]

Interdisciplinary team interaction

Interactive research has two aspects: interactions between researchers themselves; and interactions between farmers and researchers.

Interactions between researchers themselves have been a subject of study and reflection (see eg, Hildebrand, 1981; Maxwell, 1984; Rhoades et al, 1985). A major contribution of farming systems research has been the recognition that many disciplines can contribute to understanding farmers' problems and opportunities. Sometimes more concern is expressed about difficulties of interaction between people with university training in different disciplines than about interaction between farmers and university-trained researchers. Indeed, problems of 'team interaction' among the scientists of a multi-disciplinary group do often arise. For example, agriculturalists and social scientists may disagree about research agendas, even though they share common goals, but it is often important to bring to bear on farmers' problems the expertise and insights of several disciplines and for this good team interaction is needed.

Several conditions have been found to promote fruitful interaction between specialists. First, field reports are unanimous in noting how this is helped when specialists work together on the ground, in joint participation in group activities, interviews with farmers, mapping, undertaking trials and all other kinds of tasks (Colfer, 1987a and c). When researchers and extension workers are in direct physical contact with the reality of the farmers they are serving, their specialist single-disciplinary preoccupations become less dominant.

Second, multidisciplinary teams evidently work best if they are fairly small and stable in membership. One danger with large teams in the field is that members will talk and listen to each other and not to farmers. Not only does this impede learning from farmers; it is also discourteous and can make farmers feel uncomfortable and inferior.

Third, it is an advantage for differences in team members' views to be anticipated and discussed before disagreements arise. Such early discussions can then be used constructively to shed light on a subject from different angles.

Fourth, report-writing provides opportunities for interaction. Responsibilities for writing can be divided among team members both to speed up production of a report and to make sure that all contribute to the final recommendations. However, Colfer (1987a) comments that disciplinary conventions differ and each team member naturally feels some professional responsibility to abide by his/her conventions and emphasize particular topics. There may be no easy solution to the dilemma of wishing to combine the value of different insights and orientations and of diverse views, with the need for consistent style and coherent conclusions and here, as elsewhere with teamwork, negotiations and compromises are needed.

Perhaps more important than these aspects two other vital issues concerning the composition of research teams are: the balance between sexes in team membership and the role played by social scientists. These can significantly influence interactions between the researchers and the farmers.

Women's key role in agriculture is well known, as farmers, heads of households and through their responsibilities for grain storage, fuel collecting, seed gathering etc. Understanding women's roles and working with women can be difficult where the researchers are mostly men. Even a senior woman in a research team can be frustrated in efforts to encourage participation by women when local team members are all young men (Colfer, 1987c: 8). There are some topics and areas for which all-female or even all-male teams may be appropriate but research teams should usually aim for equal numbers of each sex at senior professional and extension levels. Ensuring the 'ideal' gender balance can be difficult where there is a lack of women who work professionally or as assistants in this field. Interviews with women farmers are usually best conducted by female team members who know the local language, but it is then important that the women interviewers are also involved in analysing and writing up the results.

Similarly, anthropologists or social scientists (of either sex) in teams can help foster close and effective interaction with farmers. In section 2.2, Box portrayed the social scientist as an intermediary or 'two-way translator', explaining farmers' experience to agricultural scientists and vice versa. Colfer (1987a) makes a similar point, commenting that, 'most agricultural scientists do not know how to get the information they need on farmers without some help', and adds that unless the agricultural and social scientists communicate well, the latter usually do not understand the needs of agricultural research. She concludes that there are important 'structural and ideological aspects of project management' that can either facilitate or hinder farmer involvement and these depend on whether the input of the social scientist is 'integrated or marginalized'. In a climate which is 'hospitable' to interdisciplinary communication, a social scientist can function effectively as 'spokesperson for and liaison to farmers'.

But without interest and support from team-mates, 'the social scientist is hamstrung. The valuable contribution that he/she could make is never realized. The lack of interaction keeps the social scientist ignorant of what is relevant to the agricultural scientist and keeps the agricultural scientist ignorant of what the social scientist is finding out. Titbits of information that may make their way to the agricultural scientist are then likely to be attacked as (perhaps rightfully) irrelevant – thus digging the gulf between the disciplines still deeper. Meanwhile for the farmers, it's business as usual . . .' (Colfer 1987a: 9).

The diagnostic stage in research

Preoccupations with team interactions and team management can divert attention from the primacy of farmers' priorities. The challenges of rural development now include not just raising productivity but also increasing sustainability, both economic and ecological (Conway 1987a).

The key to sustainability is that interventions help people to meet their priorities and are so fully compatible with local culture that farm people can build on them independently by means of their own experiments (Bunch 1985). People will sustain what meets their objectives and reject what does not. This requires a reversal of the one-sided relationship between specialists and farmers, so that specialists learn from farmers, with mutual learning and exchange of ideas, skills and knowledge.

Growing recognition of the need to involve farmers in research is found in countries as diverse as Thailand (Charoenwatana, 1987) and Colombia, as well as those countries which have figured prominently in previous pages: Nepal, the Philippines, India and Bangladesh. The shift of emphasis towards the realities of farming and of farm families has progressed under various labels or banners, including farming systems research (FSR), rapid rural appraisal (RRA), and now farmer participatory research (FPR) (Farrington and Martin, 1987). All of these aim to get closer to the farming realities but they do not always or necessarily involve farmers or farm families to a major degree.

In FPR for example, the farmer may participate, but the work is often 'researcher-driven' and generates insights only within the researcher's categories of thought.

While most researchers agree on the desirability of directing research toward topics which farmers perceive as problems, the way they go about identifying such topics differs greatly. At one extreme, farmers are merely observed or information is obtained without any dialogue. In other cases, information is obtained *from* farmers through surveys and questions and the research agenda takes its origin from some aspects of the farmers' situation, but it is not an agenda developed *with* or *by* farmers. It is an agenda based on questions asked by researchers and there is no guarantee that these are questions which make sense to the farmer, or that they are capable of pointing the research in a direction that will benefit him or her.

Ashby et al (1987) are among practitioners in this field who sense that an excessive concern with refinements of method is tending to divert researchers from talking interactively with farmers. They complain that: 'diagnostic research has become a hothouse of methodology development, spawning *sondeo* teams, informal surveys, rapid appraisals, key informant surveys etc. The farmer is an object of investigation, just as plants, soils, insects or viruses are objects to be studied and measured. In this process the farmer's voice has been lost. Asking farmers questions has become an industry. Listening to farmers has been forgotten as a research tool.'

In the move towards giving more prominence to farmers' agendas, many approaches and variants have been used. In Colombia, the adaptive on-farm research now seen as necessary begins with 'diagnosis of a technological, social and economic situation', in which two approaches to farmer participation are being tried. They are 'auto-diagnosis' in which small farmers define and analyse the technological realities of their farms and express their own views in their own frameworks, and 'participative diagnosis' (Chaves, 1987:15), in which cultivators express the problems they face and their needs and expectations in relation to agricultural technology.

Similarly, Eklund (1987) argues for a form of rapid rural appraisal as an improved diagnostic method, taking great care to ensure that farmers' agendas are elicited. He cites his study in Zaire which included interviews at the household level and interviews and dialogue with farmers in groups. Statistical analysis of data also shed light on farmers' practices. Open-ended questions were included to guard against imposing researchers' agendas on farmers' responses. Farmers were asked to state their three main problems, regardless of domain.

The importance of open-ended enquiry is paramount. As Box comments in section 2.2, some appraisal methods presuppose that the researchers know what questions to ask. By contrast, in an interactive mode farmers contribute to formulating the questions. That this happened in the Philippines cogon work is evident from the surprise which the researchers expressed at the topics which emerged and also by their willingness to interpret any loss

103

of interest by farmers as an indication that the wrong questions were being asked.

It is also instructive to note the sense of excitement and discovery among those investigators who do feel free enough to let research agendas develop new directions after interaction with farm people (sections 2.2, 2.5 and 2.6). An open, unreserved and completely mutual exchange is implied by 'interaction' and researchers stand to benefit as much as or more than farmers in terms of the unknown and unexpected which are brought to their attention.

Reversals in diagnosis, whatever their labels, shift initiative, analysis and choice to farmers and farm families. This implies and requires that they can command and use the tools of analysis. One part of this is to enable them, in Gordon Conway's words: 'to better analyse their existing situations so that they can understand the likely impact of interventions and innovations from outside and hence make sounder development decisions' (Conway 1987b). But beyond this, it is for them to analyse and actively generate requests and demands, and seek and use support from outside. For this, methods such as those described in preceeding chapters need further development and adoption and outsider professionals need the orientation and commitment to enable farmers to use them and then to be guided by the results.

Respect for farmers

The dimension missing from most accounts of farmer-first approaches, whether described as parts of FSR, RRA or FPR, is the basic personal attitude of the outsider professional to the farmer. Often there is an underlying conviction that the modern specialized knowledge of the outsider has a universal validity and application which should override whatever farmers know. The attitudes, demeanour and behaviour which go with this belief prevent learning from farmers. Reversals of behaviour and attitude, to respect farmers as people and to desire to learn from them, are essential complements of the farmer-first methods described in this book.

Behaviour and attitude interact. The most effective way to change attitudes – from despising or undervaluing farmers' practices and knowledge to recognizing their validity – may often be to start by changing behaviour through the adoption of farmer-first methods. One example is team composition. The value of multi-disciplinary teams walking through an area and interviewing farmers is frequently mentioned, but much depends on whether farmers are included as team members, as happens in the Philippines (section 2.4) but not so explicitly in Nepal (section 2.3). Then there is the question: if they are included, are they merely regarded as guides, or do they play an equal or dominant part with the researchers in discussion? Another example is mapping and diagramming. Mapping may be an important tool for researcher-planned work, as described by Edwards (section 2.7), in which the use of maps for agronomic monitoring facilitates group interaction among members of an outsider research team,

104

to the benefit of resource-poor farmers. Beyond that, mapping can be used in a farmer-researcher interactive mode, with cultivators, pastoralists or village-level extensionists themselves drawing maps and diagrams, or where maps and diagrams are developed through discussions with a group (sections 2.5 and 2.6). The key questions are whose knowledge and ideas determine what is represented, and whose analysis elicits priorities.

Yet another example is ethnohistorical and biographical descriptions. Asking farm people about the history of their village and its landscape and about their lives has been recommended by several authors as a means of identifying problems for research and of developing research agendas in collaboration with farmers (Okali, 1983; Rocheleau and Weber, 1987) and forces outsiders to listen and learn.

Methods such as these can be used to change not just behaviour but also attitudes and understanding. Required to include farmers in teams, to elicit maps and diagrams from farmers or to ask for ethnohistorical and biographical histories, outsiders can be helped to listen and learn. Listening and learning can in turn lead to a change of attitude from superiority to respect. This respect for farmers extends to giving them credit for information and innovations. In Part 1 of this book, it was suggested that when farmers are responsible for a technological innovation, they should be recognized and acknowledged by name, a principle which applies also to innovations in research methods.

Many of these, and other reversals, are not easy. Conventional professionalism is so strong that only resolute and sustained reversals can achieve the best balance between the knowledge, ideas and analysis of outsiders and farmers. Fortunately, as the contributions to this book show, methods for these reversals exist, and are being further developed. The means do exist for identifying farmers' agendas and for putting them first. For them to be well used requires of outsiders a transparent respect for farmers, a sensitive interaction with them and a recognition and acceptance of them as fellow professionals and colleagues.

Notes

1 This paragraph comes from Gupta (1987a), p. 52.
2 Based on Gupta (1987a), p. 58–61.
3 Based on discussions in the ITK study group and informal comments by Anil Gupta, Roland Bunch, Lori Ann Thrupp and Ed Barrow.

PART 3

Practical participation

Introduction

The theme of Part 3 is practical ways in which farmers can participate in agricultural research, especially research on-farm. James Sumberg and Christine Okali discuss the rationale and types of on-farm research, arguing that the farmers' role in technology development becomes more critical and cost-effective the more complex the technology, with alley farming in Nigeria as an example. Jacqueline Ashby, Carlos Quiros and Yolanda Rivers describe how in Colombia farmers' opinions on varieties of bush beans and cassava were elicited, farmer-managed on-farm trials were conducted and evaluation was based on farmers' preferences.

Groups and workshops, a prominent part of farmer participatory research, are treated in three sections. Authors describe how farmer group meetings can contribute to variety selection and to agroforestry and technology assessment and how they can be used to elicit information, exchange ideas, determine priorities, plan trials and monitor and evaluate results. The forms and contexts of group meetings include field hearings about livestock in north-east Brazil (Baker and Knipscheer), pastoralism in north-west Kenya (Barrow), the evaluation of crop varieties in Colombia (Ashby, Quiros and Rivers) and field days in Zambia (Kean, Edwards). Workshops of innovator farmers are described by Ashby, Quiros and Rivers for crop variety selection and by Zainul Abedin and Fazlul Haque for scientists to interact with and learn from farmers in Bangladesh. From Botswana, David Norman and his colleagues present a typology of groups and their strengths and shortcomings in the process of technology development.

Finally, Maria Fernandez and Hugo Salvatierra describe participatory technology validation, including modifications of trials, in highland communities in Peru and Carol J Pierce Colfer with her colleagues contrasts different methods for farmer involvement in research in Indonesia.

As these and other examples indicate, farmer participatory research can take many forms, has already an extensive repertoire of methods and has proven itself both practical and cost-effective.

3.1 Farmers, on-farm research and new technology

J SUMBERG AND C OKALI

Approaches to on-farm research

Over the last two decades, a growing interest in farming systems research has highlighted the potential importance of 'on-farm research'. However, research activities on farms have taken many forms. Some researchers have essentially transplanted research objectives and methodologies from the experiment station to the farmer's field, while others have attempted a more whole-hearted incorporation of the farm and farm family into the research activity. Most commonly, on-farm research is seen in terms of validation and demonstration of technologies which have previously been developed elsewhere under controlled experimental conditions.

On-farm research must surely encompass a range of approaches and activities and has a role to play in all stages of agricultural development, from the identification of new technology to its validation and demonstration. The inclusion of farmers in the process of technology development has, however, been relatively rare. It is this aspect of on-farm research that we will address here, drawing from experience in West Africa with the agroforestry system known as 'alley farming'.

The farming systems approach to agricultural research is predicated on an appreciation of the whole farm enterprise and, most importantly, of the relations among its component parts.

Incorporating this 'holistic' view of the farm and farm family into specific research strategies has not been straightforward. One fairly drastic approach has been to establish farm families on 'model' farms located either on experiment stations or within farming communities and thus provide researchers with an opportunity for detailed study of new farming technology (Menz, 1980; Collinson, 1982). Perhaps the more common approach, however, has been through the use of research on farmers' fields, which has taken forms ranging from standard crop variety trials to detailed studies of availability and use of family resources.

In much of the on-farm research with new crop varieties and alternative production practices, the farmers' role has been to provide land and labour, act as an experimental control by farming an adjacent plot with his or her 'standard practices' and, finally, to react to the results of the experimental treatments.

How did on-farm research become synonymous with on-farm trials used only for this sort of validation and demonstration exercise, leaving farmers with no role in innovation? A partial explanation can be found in the emphasis which has been placed on the development and dissemination of new crop varieties. Plant breeding and variety development activities take place predominantly on research stations. Except for the key role farmers can play in identifying breeding and improvement objectives – demonstrated in section 3.2 below – there are few well developed models for farmer participation in the early stages of crossing, selection and evaluation. Once potential new varieties have been identified and stabilized, often with a strong emphasis on yield and productivity, on-farm trials are used to determine how a limited number of these perform under conditions which are presumed to be more 'realistic' than those on the experiment station. The new variety may or may not be acceptable: in any case the farmers' input is often highly structured (in order to obtain comparable data from a number of trials) and only enters in the final stages of evaluation. While it may be that farmers have only a small contribution to make during the very

110

early stages of variety development, it can be argued that highly structured on-farm trials limit the farmers' ability to experiment with and manipulate the new genetic material, thus precluding potential adjustments in other production practices or exploration of production niches which might make the new variety more interesting (Richards, 1985).

Another part of the answer to this question can be found in the common agricultural research and development strategy which stresses the importance of 'technical packages' as opposed to changes in individual management components. These packages, which in the case of crop production might include a new variety, higher plant density or altered planting arrangement, fertilizer and pesticide recommendations, are often an all-or-nothing proposition: the individual innovations, taken alone, may contribute little within the context of the existing production system. In this case, the objective of on-farm research is not to identify opportunities within the current production system, nor to develop or evaluate individual technical innovations. Rather, the objective is to determine whether experiment station results are reflected in the package's on-farm performance and, if not, to identify the constraints relevant to the closing of this gap.

Parameters, variables and constraints

The identification of constraints in existing production systems and of constraints on the performance of packages of new, 'improved' technology, are key objectives of farming systems research. We would argue that the second activity – identifying constraints on technology packages – predominates. On-farm trials are viewed as a means of validating a given package and the role of the farmer is to highlight, through his or her own actions, these constraints, which have generally been identified as tradition and attitudes, institutions, knowledge, input availability and credit (Gomez, 1977). In this scenario, everything which is seen as limiting the package's performance is a constraint or a variable which can and should be addressed. The whole system is therefore open to question and, given these assumptions, is justifiably subject to manipulation. Much of the subsequent research or on-farm activity revolves around removing these constraints in order to close the gap between the package's performance on the experiment station and on farms. It is in this context that we can better understand the moves by some farming systems research programmes to provide cooperating farmers with inputs and credit and even to engage in manipulation of the marketing system.

An alternative approach to on-farm research is to begin with the system itself, which inevitably means recognizing the complexity of smallholder farming – its multi-purpose objectives and the complementary nature of the system's parts. A large commercial farm usually has fairly simple objectives, such as producing more grain for a given level of costs and other goals related to market opportunities. With small farms, by contrast, objectives are related much more closely to the family (and perhaps other people) sustained by the farm. Simply because the farm is smaller, people are a bigger part of the picture and so are the many aspects of their needs and activities. It is the complexity which makes the problems of small farmers and resource-poor farmers so difficult to tackle by working solely on experiment stations and producing 'packages'.

We contend that the farmers' role in technology development becomes more critical and increasingly cost effective as the proposed technology becomes more multi-faceted and complex. In these circumstances, classical methods for designing, refining and evaluating technical innovations become less useful. A good example would be the conceptual and experimental pitfalls inherent in research on even relatively simply inter-cropping systems. As we look to even more complex technologies such as agroforestry systems which can potentially produce crops, wood, fruit and fodder, it is obvious that a traditional experimental approach seeking to identify management treatments which maximize an output becomes unwieldy and unrealistic. It is the farmers themselves who hold the keys for developing, evaluating and validating these systems (Okali and Sumberg, 1986; Atta-Krah and Francis, 1987).

Alley farming is a new, multi-faceted technology that can potentially be used to fulfill a wide range of objectives for resource-poor farmers. Its recent development illustrates the limitations of experiment station research and the possibilities for on-farm research. Alley farming is a method of crop production within 'alleys' formed by rows of fast-growing trees. The trees are pruned and managed primarily to benefit the crops. The technique is essentially a modified and intensified bush fallow approach to soil fertility, with the fallow vegetation being replaced in time and space by the rows of fast-growing, often leguminous trees. The fundamental assumption is that by inter-planting the trees (the fallow) with the crops and by pruning the trees to produce a fairly constant supply of nutrient-rich foliage for application to the soil, continuous cropping at a reasonable yield level can be sustained (Kang, Wilson and Sipkens, 1981; Kang, Wilson and Lawson, 1986).

Research on alley farming

Experiment station research on alley farming has demonstrated that the basic assumptions which underlie the technique can be valid in a number of humid and sub-humid environments and with a range of crops. Moreover, alley farming systems are in use in parts of Asia, where rows of trees on steep slopes play an additional role by helping to stabilize the soil. After work in south-west Nigeria by IITA and ILCA, alley farming is being considered as the basis of a government development programme for small livestock producers (Sumberg and Okali, 1984). Thus on-farm research is related to the opportunities presented by this programme.

When alley farming is the subject of work on experiment stations, issues such as height and frequency of tree pruning tend to loom large. However, once alley farming is put into the hands of farmers, such matters cease to be important research themes. Farmers will prune the trees at a height which is most comfortable for them and which minimizes shading on the accompanying crop. Experiment stations may find that an alley-width of

112

2 m is optimum, but on-farm experience in West Africa shows clearly that this is unacceptable, regardless of considerable experimental data concerning tree foliage and crop yields. Farmers considered two metre alleys too narrow to work in comfortably and if tree pruning was delayed only slightly, the potential damage to the crop from shading was considerable (Ngambeki and Wilson, undated). Thus, with some sensitivity and just minimal farmer input, it is possible to draw the general outlines of an alley farming system – a working model – which can be taken to a larger group of farmers for development and further refinement.

We do not mean to imply that there is no role for more traditional research approaches. In fact, before alley farming can be taken to the farmers, even on an experimental basis, questions such as appropriate tree species and reliable, economical establishment methods must certainly be addressed. It can also be anticipated that new research questions will arise from the farmers' experiences and that some of these may be most appropriately explored on highly controlled research plots (Norman and Collinson, 1985). It is this dynamic interplay between station and on-farm research than can help realize the potential of the farming systems approach.

The key research objective – for both the farmers and researchers – is to develop a broad understanding of the range of alley farming management options and the ways these options can be used to fulfil a variety of objectives. Two kinds of issues need to be explored.

The first relates to the conditions under which the basic components of the technology can be made to work. Alley farming, for example, is based on the presence and use of fast-growing trees and it is therefore important to understand the conditions under which the trees can be successfully established (including factors such as soil, establishment year, crop combination, time of planting, etc). The objective is not to identify an optimal or recommended set of conditions for tree establishment, but rather to develop an appreciation of the flexibility in management of this particular component. One of us (JS) was appalled and discouraged in 1983 when some farmers participating in ILCA's on-farm research in Nigeria ignored our recommendations and chose to establish alley farms by planting trees with highly competitive crops such as yam, melon and cassava instead of maize (with which we had experience and which we assumed would be less competitive and therefore favour early growth and establishment of the tree seedlings). Much to our surprise the tree seedlings survived, in spite of being engulfed by yam and melon vines or shaded by cassava and, while the establishment of the trees may not have been as rapid as with maize, the lesson was very clear. The technology was being made to work within the diversity which characterized the local production system. The major objective was to get the trees into the system and we eventually realized that it was somewhat inconsequential if it took 6, 9, 12 or even 24 months for the trees to become established. As a result of experiences such as this, our view of the potential and flexibility of alley farming was greatly expanded.

The second set of issues relates to the conditions under which the technology as a whole becomes interesting or valuable to farmers. For example, farmers in ILCA's on-farm research activities in Nigeria demonstrated that a much wider variety of crops and crop combinations could be successfully grown in alleys than had previously been tested on the research station. The on-farm research also afforded an opportunity to observe how the technology was used within the context of the farm and farm family, for example, by highlighting issues related to the decision-making processes governing the use of tree foliage for crop production by men and/or small ruminant feed by women (Okali and Sumberg, 1986).

On-farm trials or on-farm research

What form should on-farm research take if it has as its objective farmers' participation in technology development? At this point we want to make a distinction between on-farm research and on-farm trials. The basis of a trial – whether it takes place on-farm or on an experiment station – is to compare two or more options: it is obvious that the alternative technologies must, therefore, be in a relatively well-defined state. Since the thrust of our argument is that farmers must be incorporated into the process of development and that the purpose of the on-farm research is to provide farmers (and researchers) with an appreciation of the options presented by the technology, it should be clear that on-farm trials as they are most commonly structured in farming systems research will be of little value.

The fact that on-farm research might exclude conventional on-farm trials, whose principal objective is the validation of packaged technology, does not mean that there is no need for evaluation in the research process. Indeed, another important distinction between standard on-farm trials and on-farm research that aims at technology development is the kind of outcome variables that are of interest for evaluation. It is probably fair to say that the central outcome variable in most on-farm trials is crop yield. Other variables may include such things as labour inputs, crop quality and farmer satisfaction. However, since the purpose of the on-farm research we are discussing is to develop an appreciation and understanding rather than pick a winner, the question of appropriate outcome variables becomes crucial. We submit that the most appropriate outcome is the farmers' interest, which can be assessed via a number of questions: is the plot being farmed in the second year? Is the alley farm being enlarged? Are neighbouring farmers planting alleys? Are the farmers developing new ways of using the technique?

On whatever basis the farmers are making these decisions, one can be reasonably well assured that it involves a level of analysis and synthesis which goes far beyond even the best formal experiments designed to 'provide a valid assessment under farmer conditions' (Farrington and Martin, 1987). The need is *not* to keep the trials 'simple enough for farmers to understand and evaluate' (ibid) nor to develop more sophisticated statistical methods, but rather for research and research institutions to accept the proposition that *adoption by farmers is validation of a technology*, one might say by definition, even if we are unable always to identify or quantify the technology's effects.

114

3.2 Farmer participation in technology development: work with crop varieties

JACQUELINE A ASHBY, CARLOS A QUIROS, YOLANDA M RIVERS

Farmers at first hand

Technology development programmes oriented to small farmers often start with elaborate surveys designed to set objectives for the on-farm experiments and to formulate research agendas. The complexity of small-farmers' decision-making is such that it can take a team of specially trained researchers weeks of fieldwork to achieve this, using *sondeo* teams, informal surveys, rapid appraisals, key informant surveys, etc. The assumption behind this activity is that farmers cannot articulate their problems and goals but require highly educated intermediaries to interpret their preferences. There is, of course, some truth in this, but it has seemed to us worth testing the alternative view that in crop breeding programmes, at least, farmers can articulate their views directly to researchers. Therefore, the farmer participation in trials of crop varieties discussed here was designed to address the following questions:

- instead of using surveys to interpret farmer preferences and objectives second hand, was it possible to elicit these preferences first hand, by structuring appropriate situations where farmers could and would express their criteria for selecting among different varieties?
- could these preferences and criteria then be applied to selecting varieties suitable for testing within the farmers' own cropping system?

The farmer evaluations which followed were conducted in 1985 at Pescador in Colombia at an early stage in on-farm research. The objective was to pre-screen with farmers a large number of improved varieties of bush beans and cassava; to determine which were worth testing in farmer-managed trials; to obtain suggestions from farmers on how factors other than variety, such as fertilization or planting distance, should be incorporated into the design of on-farm trials. We were therefore attempting to involve the farmers in the design of the trials, as well as carrying them out, which is very rarely done.

Farmer evaluations have been carried out in several types of trials, such as regional trials which include large numbers of varieties (as many as 35 varieties in one case); exploratory trials with up to 10 varieties at two different levels of fertilization; farmer-managed trials with up to 10 varieties superimposed on farmers' levels of management in the test crop.

The project has therefore consistently worked with relatively large numbers of treatments for farmer evaluation. This has been done in an effort to test another assumption common in on-farm research, that farmers can only evaluate limited comparisons such as, for example, the researchers' 'best' treatment compared with a farmer check.

115

Pre-screening varietal materials by farmers

The first step in establishing trials of crop varieties was to give farmers an opportunity to select materials from a wide range of possible varieties for inclusion in the trials. With bush beans, individual farmers were shown samples of the seed from different lines identified as promising for their agroclimatic region by the CIAT bean programme. Each farmer was asked to indicate those grain types of interest and those less acceptable. With cassava, a group of farmers visited a CIAT regional varietal trial at a site near the research area where the cassava plants on the borders of each experimental parcel in the trial had been left standing. Farmers were therefore able to inspect plants of each variety, uproot sample plants to examine cassava roots and thereby make a selection of the varieties they perceived as interesting for further testing. In a group discussion of how to test the varieties they had selected, the farmers talked about their observations that the same cassava variety would give very different yield and root quality in different fertility conditions, and suggested different fertilizer treatments for inclusion in the on-farm trials.

Prescreening of beans focused on the initial evaluation of bean grain types by farmers. A CIAT breeder selected ten bush bean materials which were potentially adapted to the agroclimatic conditions of the research site and ranked them in order of expected acceptability to farmers. Subsequently farmers examined and ranked samples of each variety and discussed acceptability with the research staff.

Farmers were readily disposed to rank the materials according to grain type: their ordering varied somewhat from that of the breeder, because their most important criterion for grain acceptability was grain *size*, as shown in Table 3.1. There was however, one intriguing exception to this rule: the interest shown in a small grain variety, BAT 1297. Analysis of the interviews in which farmers made these initial selections suggested that the unexpectedly high ranking given to this variety was the result of women taking part in the selections and their perceptions that traditionally small grain varieties similar in appearance to BAT 1297 had been the more flavourful and higher yielding. Women viewed a small grain type such as BAT 1297 as desirable from the point of view of subsistence and consumption objectives of the small farm. Men on the other hand were selecting grain types for size primarily with reference to marketability.

The initial selection of varieties was intended to accomplish two objectives: first to ensure that materials obviously unacceptable to farmers did not enter the on-farm trials; second to create an opportunity for farmers to make suggestions about how trials to be established on their farms should be designed. This helped to ensure that the crop varieties would be tested in conditions that farmers viewed as realistic and representative for their conditions. It also established in the farmers' minds from the outset, that the trials were not intended to convince them that any given variety was superior. This is important because small farmers who have encountered scientific experts before expect that the experts will tell them what to do, and try to convince them that 'new' is 'better'. In order to create a relationship with farmers where free and open communication

116

Table 3.1: Prescreening Seed According to Grain Quality

Bush Bean	Grain Type	Farmers' ranking[1]	Breeders' ranking
AFR-205	Large, purple mottled	1	3
A-486	Large, pink opaque	2	2
A-36	Medium, red opaque	3	1
ANCASH-66	Medium, white	4	9
PVAD-1261	Medium, white	5	7
BAT-1297	Very small, red opaque	6	10
G-4453 × BAT 1386 C	Small, red opaque	7	8
HORSEHEAD XYC 206	Small, red opaque	8	4
G 7223 × BAT 1276 C	Small, red opaque	9	6
ANTIOQUIA 8L-40	Small, red opaque	10	5

[1] Farmers were asked to select six preferred grain types out of the total ten and rank them from most preferred (score = 6) to least preferred (score = 1); the final ranking is based on total score for each variety.

about the performance of new technology in their fields is the norm, research staff need to convey to farmers the importance of expressing opinions frankly about what is acceptable.

The farmers who were invited to take part were identified by asking local farmers to name people whom they considered 'expert' bean or cassava producers, defined as years of experience and interest in trying new ideas for cultivating these crops. Each farmer interviewed gave opinions as to who were the local experts in one or the other crop. The additional farmers identified by this approach were also visited, with the result that the list of names expanded and certain individuals were repeatedly mentioned. Those farmers whose names occurred more than twice were considered experienced cassava farmers or expert bean growers in the community and were invited to the pre-screening exercise.

In the trials which followed, farmers had to plant a commercial-scale plot of the test crop, so that the trial was situated within the plot. The nine farmers who took part in each set of trials were chosen from the lists of bean or cassava growers compiled for the pre-screening stage, but were selected carefully to represent a range of socio-economic resources. This was done by observation of their housing quality, ownership of consumer durables, ownership of livestock and other qualitative indicators.

In each subsequent season when trials were planted, the farmers who managed them were different individuals from the previous season. The project wanted to avoid the 'trained farmer' syndrome, so that fresh evaluations could be obtained in different seasons with the same varietal materials and also to avoid pestering farmers with repetitive interviewing.

Trial establishment

For testing bean varieties the trial design consisted of eight bean varieties including the local variety as a check, superimposed on farmer management practices which varied from farmer to farmer. Each of the 15 farms on which this trial was planted was a replication. Farmers designated sites for a varietal trial within a field where they planned to plant beans. Farmers planted one bean variety to each of eight plots staked out at the designated sites, some ordering their varieties by grain size, others by grain colour. The remainder of the bean field was planted by the farmer with the local variety. Planting densities, fertilization and all other crop management operations in the trials were determined and carried out independently by each farmer.

The cassava trials used two approaches. The first was similar to that used in the bean variety trials, but because of a shortage of planting material and the desire of farmers to include fertilizer treatments in the trials, it was not possible to give each participant all the varieties to evaluate. Instead, each of nine farmers was randomly assigned three of the CIAT varieties and the local check to be planted at two levels of fertilization, as defined with the group of farmers in the initial selection discussed above. Since the main objective of these trials was to obtain farmers' reactions to the new technology, it was decided to reconvene the same group of farmers at harvest time to pool experience and to see if a group could produce a consensus about which materials looked sufficiently promising for them to be included in further on-farm testing.

The second approach (suggested by Ted Carey, Breeder, CIAT Cassava Programme) was simply to make gifts to farmers of three cassava varieties in separate packages of planting material, each labelled with a different number. The farmer was told that he or she could plant the material however desired and that follow-up visits would be made by research staff to see how the varieties were performing. Follow-up visits suggest that a few farmers may be planting the cassava stakes in a way that reflects concern that the agronomist might be displeased if the gift is not given special treatment. In these instances, the purpose of the exercise is being frustrated in that researchers are not able to observe an authentic farmer reaction to the material.

Evaluating varietal trials with farmer participation

The central objective of establishing the varietal trials was to create opportunities to listen to farmers' reactions. The primary data of interest in trial evaluations are therefore the opinions, preferences and ideas expressed by farmers.

These 'preference evaluations' have been conducted on two occasions in the growth cycle of each of the two crops. In the bush bean varietal trials, one evaluation is carried out when the bean pods first begin to form. In the cassava trials, the first evaluation is carried out after farmers have carried out the first weeding. These evaluations focus on plant architecture, growth habits, disease susceptibility, periodicity and aspects of management to date, as observed by farmers. The second 'preference evaluation' follows the harvest and focuses on yield, profitability, marketability and consumption aspects, but aims to capture any relevant varietal characteristic which farmers like or dislike.

The 'preference evaluations' require interviewers skilled in the techniques of open-ended interviewing, which involves stimulating the farmer-respondent to express opinions and concepts, and to explain observations of different varieties, without prompting or suggesting that the interviewer has a point of view, which will bias farmers' responses. Experience with these interviews suggests that it is helpful for the interviewer to communicate a priori a complete absence of vested interest in the 'success' of the new varieties relative to local varieties. It is necessary to bear in mind that farmers in these interviews tend to expect that the research staff want to hear favourable comments. The interviewer has therefore to establish at the outset that a farmer's negative reactions are equally acceptable, and that the reasons for such reactions are of profound interest, without prompting the farmer to express ideas that appear to be what the interviewer wants to hear (Table 3.2).

At the time of trial establishment each farmer is given a map of the trial, stored in a plastic folder, with a simple form in which labour and other inputs to the whole field are recorded by the farmer.

When visiting a trial, the research staff make a point of relying on the farmer to show them around. The aim here is to communicate the feeling that this is the farmer's trial for which he or she is responsible, not the researchers' trial. Reliance on the farmer to act as the guide around the trial layout also enables the research staff to assess readily how seriously the farmer takes the trial: whether or not he or she knows his way around the trial and can identify the different plots and varieties is a sure indicator of how attentively the farmer is observing their progress. Most farmers readily locate varieties they are interested in without using a map, and have evidently studied them because they learn the reference letters and numbers of varieties (such as A-36, XAN-212, BAT 1297) and can locate them in the trial by these names, or by the numbers given to the cassava varieties (51, 52, 53, etc). In several instances farmers have begun to name cassava or bean varieties of particular interest to them from their appearance, eg, 'la blanca' (cassava hybrid CG 406–6) or 'la pequenita' (bean variety BAT 1297).

To date the most workable procedure for carrying out preference evaluations in the trials has been to note farmers' comments in columns under broad topic headings in the order that these are spontaneously brought up by farmers during the interview. At the conclusion of the interview the farmer is asked to indicate the three or four varieties of most interest, rank these in order from most to least interesting, and explain this ranking.

119

Table 3.2: Excerpt from a farmer's evaluation of the standing bean crop, Pescador, Cauca, 1985

INTERVIEWER: Which variety is this? What's this one called?
FARMER: A=66, I don't like this plot very much. This one has a lot of leaves and few pods.

INTERVIEWER: A lot of leaves?
FARMER: And few pods, that's to say that it's flowered a lot but only formed a few pods.

INTERVIEWER: Is it a disadvantage that it has a lot of leaves?
FARMER: Yes, it really has a lot, I'm afraid of a bean plant which is so leafy, it's grown into a mountain of bean plants and then the diseases take hold more easily because of the humidity.

INTERVIEWER: I see.
FARMER: That's why I don't like it, because it has to be planted a bit further apart and as it's formed few pods well, the yield would be very low. So it's better to look for a variety which doesn't have a lot of leaves and which has more pods.

INTERVIEWER: That's ve y important for us to know ...
FARMER: There's no comparison with another plot we can see over there, a plot I really like, it's formed a lot of pods and the plant is like this ... look.

INTERVIEWER: It's small?
FARMER: Yes, look at the number of pods, you can tell from a distance. So that plot will yield more because you can plant closer and you can harvest more (plants). Now this here is very nice plot, I like this bean a lot ... Look at the difference between this bean plant and the one we saw over there with lots of leaves. The plant is still pretty leafy but it also formed a lot of pods, look at this plant here ... it's healthy—almost no disease. Yes, this plot is really healthy-looking. Do you know what to do with this type of bean plant, they're bush beans but they put out tendrils, the right thing to do for this one is always plant in a straight line so you have space to walk through between the lines.

Subsequently it is possible to do a content analysis of the interviews recorded in this fashion and to tabulate the frequency with which specific varietal characteristics or criteria for evaluating the crop in question have been spontaneously mentioned by farmers. Table 3.3 is an example of a frequency tabulation of criteria mentioned by farmers in bean varietal evaluations. Performing a content analysis with the frequency tabulation highlights the criteria which are important considerations for farmers in a visual evaluation of the bean crop: for example, farmers universally commented on yield potential. While several noticed disease resistance or susceptibility, this did not appear as an important criterion for farmers to rank the varieties placed in first, second and third place in Table 3.3.

The second preference evaluation which follows harvest of a trial involves weighing grain or root yield from each experimental parcel which is recorded with the farmer on a prepared form. The farmer and agronomist evaluating the trial each make a copy when farmer literacy permits. Yields are expressed in returns to seed or to area, depending on local units of measurement for expressing yield for the crop. The farmer is asked to rank treatments in order of preference based on a visual appreciation of yields and quality aspects of different varieties. Next a simple cost-benefit analysis is performed, calculating value of the yield according to the prices obtained by each farmer for each variety and costs of purchased inputs of concern to the farmer. The farmer is then asked to rank treatments in order of preference based on this cost-benefit analysis and the reasons for preferences are discussed.

Table 3.3: Farmer Evaluations of the Standing Crop (9 farmers)

VARIETY	BEAN VARIETY CHARACTERISTIC							
	High yield	Low yield	Early	Late	Disease resistant	Infected	Upright plant	Sprawling plant
Rank *Most preferred*								
1 A-486	9	–	2	–	1	2	1	–
2 PVAD-1261	9	–	–	1	2	1	4	–
3 BAT-1297	7	–	–	1	2	–	–	4
4 A-36	5	–	–	1	1	1	–	1
Least preferred								
Calima	5	–	–	–	1	4	–	–
Antioquia	–	1	–	5	1	3	–	1
Ancash-66	–	1	–	5	3	1	–	4
AFR-205	–	1	–	–	1	4	–	1

Sometimes a simple visual appreciation is the most effective evaluation, but often the cost-benefit analysis is a useful tool for eliciting perceptions of constraints when costs or prices vary among treatments in a trial.

Concluding remarks

The value of these farmer evaluations is not principally to select the two or three varieties which the majority prefer, but more important, to understand the objectives which farmers are addressing as they make selections among crop varieties. This information is important for narrowing down the combinations of varietal characteristics which breeders consider when taking farmers' views into account during breeding programmes.

Following the on-farm trials, the project undertook further research into farmers' preferences, using group participatory methods, which are explained as part of section 3.4.

3.3 Farmers' groups and workshops

IDS WORKSHOP[1]

Overview of group methods: types, purposes and features

Group methods are increasingly used in agricultural research, extension, and other development activities (Kumar 1987). In Parts 1 and 2 (see 'groups' in index) we saw how group meetings can help in eliciting farmers' ideas and problems, in discovering and enhancing their knowledge and innovations and in trials and technology development. This section and those to follow (3.4, 3.5 and 3.6) present further experience.

'Farmer groups' refers to groups composed mainly of members of the rural community, along with one or more agricultural researchers and/or extensionists. There are many types, sizes and purposes of groups in agricultural research activities and developing a typology is difficult. Most groups have several functions. 'Workshop' is used for some groups usually to indicate one-off problem-solving or teaching/learning activity. 'Innovator workshops' (sections 3.4 and 3.5) usually involve a meeting to allow experimenting farmers to discuss their innovations with each other and with researchers.

Purposes of farmer groups can include:

- *building interaction and communication* between researchers and farmers, eliciting and exchanging information from farmer to farmer, from farmer to researcher and from researcher to farmer;
- *analysis* by farmers, with researcher support, of their problems and needs, reinforcing and fostering their own knowledge and capability;
- *R&D*, with the choice, design, conduct, monitoring and evaluation of experiments;
- *extension* from farmer-to-farmer, and the diffusion of innovations;
- *empowerment*, enabling farmers to organize for action or to share a resource.

Usually, any group serves several of these purposes. Some may be implicit functions rather than explicitly-stated aims. Nor are the purposes of a group static and inflexible; new aims and issues may emerge over time during the course of group meetings, especially as members desire changes. Seldom is it effective for researchers to establish rigid preconceived purposes prior to beginning group meetings.

Groups can serve as research/development methods in themselves or can be part of others. For example, diagramming, manual discriminant analysis, community appraisal, mapping and 'chains of interviews', can use groups.

Several issues related to the setting up and functioning of effective groups can be noted. Deciding on the appropriate size, membership and selection procedure also deserves careful consideration. One of the concerns is ensuring equality of composition and of dialogue, to promote constructive activities in which all group members feel free to participate and to avoid exclusion and jealousy of other community members. It is usually desirable to work with groups that are already established in an area, if they exist, as long as they have appropriate equitable composition and the group members are interested in participating in the new project. Obviously, timing and location of meetings should be planned mainly for the convenience of farmers, to ensure their full participation. Some groups can be effective for temporary activities; for instance, a single meeting may be sufficient to pass on ideas from farmers to researchers or to provide a forum for exchange of information, as described below in the cases of innovator workshops in Colombia and Bangladesh. Other groups convened at intervals can provide continuity in, for example, monitoring trials, discussing problems and progress, or carrying out self-sustaining project work. These and other points are illustrated in the following descriptions.

Groups in field hearings

Baker and Knipscheer (1987) describe groups participating in 'field hearings' in north-east Brazil. The conception was the use of farmer groups to evaluate and screen new technologies. The term 'hearings' was used to emphasize the importance of listening on the part of researchers and extensionists. (See also Knipscheer and Suradisastra, 1986 for regular research field hearings in Indonesia).

This work was carried out on a resettlement site where all farmers had the same amount of land with similar mixed crop and livestock enterprises. There were three project areas with 66 farmers in each. In one area, the researchers merely monitored the growth of farm livestock without making any other intervention. In another area, they provided a 'package' of veterinary interventions, but did not hold meetings or promote any group activities. In the third area, however, the same veterinary 'package' was accompanied by regular discussions between extensionists and farmers at 'field hearings', at which research or extension specialists also offered training in animal health, breeding or management.

As expected, the livestock in these latter areas did better than in the other two, as was shown by the rate at which they gained weight. Farmers' attitudes were also very positive in the area where field hearings were held and they showed greater willingness to pay for veterinary services.

Baker and Knipscheer conclude that the field hearings helped not only to inform farmers about the new technology, but also with farmer motivation. The farmers also 'provided important information and insights in the identification of the most limiting production constraints, and in the early stages of the project were instrumental in the choice of technologies which were tested' (Baker and Knipscheer, 1987:10).

One issue which needs to be considered when embarking on research with farmers' groups is whether to form new groups especially for research purposes, or whether to work with established groups. One consideration is whether there are appropriate common interests, and an equality of dialogue in the existing group. Where community leaders and the better-off farmers dominate meetings, it will usually be better to set up separate groups for non-leaders, women and/or resource-poor farmers. In some societies, however, it is unlikely to be politic just to form groups for the disadvantaged and it will be necessary for the needs and interests of better-off farmers to be addressed to some extent as well. Sometimes the necessary arrangements can be made by running some groups informally. For example, while men meet in a formal and official group, it may be possible to involve women in an informal group which meets at the same time.

In work with pastoralists rather than arable farmers, there may be few opportunities to use group methods for research except within the format of existing meetings. As Barrow (1987:6) comments in relation to Kenya, in the higher potential lands where people are settled and have title deeds, 'it is relatively easy to find and talk to farmers. But how is this done in the pastoral areas where people have to be mobile?'

There is still a tendency to associate pastoralism with random wanderings, yet pastoralism and rangeland utilization is anything but random. The stock owners regularly meet to discuss grazing patterns, diseases, water access and utilization, the necessity for movement to better pastures and so on. This is an ideal starting point for discussion and learning since the people involved will usually be the leaders and elders. The place where they meet is usually a focal point in the area and is often centred near a watering point under a large shady tree (Barrow 1985).

Discussion at such a forum, sensitively approached, can give researchers a good insight into the people's knowledge of their area and a perspective on their problems and aspirations. However, Barrow (1987) warns that care is needed in interpreting what is discussed. Issues such as schools, cattle dips and wells may be mentioned merely because they are topical and 'real issues' will only emerge after the researcher has been accepted by the group. It is also important to hold discussions with other groups throughout a region, because 'problems and what people know will vary in different areas and with the different groups of people (in particular the men and women)'.

124

Farmers' groups and field days

One particularly fruitful form of group activity connected with on-farm research is the farmer field-day, usually conducted in or near fields where trials are in progress. Sometimes the aim is for farmers with trials on their own land to make comparisons with similar trials on somebody else's so that problems can be discussed. Sometimes the field-day is an occasion for researchers to listen as farmers evaluate the crop varieties under trial during the growth season (Ashby et al, section 3.4). Sometimes a promising technology has been introduced and the field-day is a stimulus for farmer-to-farmer extension (Norman et al, 3.6).

Discussions of farmers' groups in two different provinces of Zambia, both with some emphasis on field-days, have been provided independently by Kean (1987) and Edwards (1987b). Both describe evolving situations in which researcher-managed trials on farmers' fields have gradually become more interactive and responsive to farmers' views. The extent of evolution in terms of the size of the on-farm programme can be seen from figures for Luapula Province, where 17 farmers participated in only four trials in 1982–3, but where numbers had increased to 60 farmers and 13 trials in 1985–6 (Kean, 1987:5).

In both provinces, the selection of farmers to participate in on-farm trials has been a major issue. In Luapula Province, it is felt that after five years, 'the team has still not found the best method of farmer selection' and there has been a tendency to end up with 'relatively more wealthy, male, progressive farmers' (ibid:12). In both provinces, farmers who are selected tend to be clustered geographically in relatively small areas. In Lusaka Province, transport difficulties have made it necessary to select farmers whose fields are within walking distance of trial assistants' homes. However, within these fairly tight clusters, it is possible to select a representative group by recruiting farmers according to criteria concerning 'access to resources' and gender. 'It was intended that both those with and without easy access to resources would be recruited' and a percentage of the farm households selected are female-headed, 'based on their actual representation within the community' (Edwards, 1987b:6). After a selection based on these criteria has been made, the soil scientist checks that the resulting pattern is adequately representative of different soil types.

There has also been an evolution in the type of trial planned in both provinces. In Luapula, all trials were originally described as 'researcher-managed', but now about a quarter are 'farmer-managed' (Kean, 1987:5). Maize and cassava varieties have both been the subject of trials and so has the cultivation of the areas with persistent soil moisture in the dry season known as dambos. Kean comments on a 'maize/bean intercrop trial' carried out because of its possible relevance to improving the nutritional status of small-scale commercial farmers. The idea arose from discussions in which commercial farmers reported that they were no longer intercropping, whilst subsistence farmers still continued this practice.

The significance of farmers' field-days becomes clear when it is realized that they initially provided the principal opportunities for farmers involved in these Zambian on-farm trials to meet as a group. In addition, they are

125

occasions when farmers meet scientists from the experiment stations. When on-farm research began, the occasional field-day seems to have been almost the only meeting held for participant farmers, but after two seasons, an end-of-season meeting was introduced in Luapula Province to discuss the results of trials and to elicit farmers suggestions for the next year's programme (ibid:8).

Explaining how field-days have evolved, Kean says that during the first two seasons, the researchers made most of the arrangements and did most of the talking and 'the farmers felt rather intimidated'. In the third year, the trials assistants did all the explaining while the researchers stayed in the background, but the farmers still did not participate very actively. Further changes then included the holding of smaller, local field-days at which the farmers themselves explained the trials which had taken place on their land. Then a second, larger meeting was held in a primary school, attended by community leaders and other farmers. The aim here was 'to encourage farmers from the different groups to voice their opinions about the trials (and) about how the trials could be improved'. The local extension officer made a record of the discussions and researchers were pleased by the level of participation of farmers and the useful comments received (*ibid*:7–8).

In Lusaka Province, field-days also evolved in the direction of smaller, local meetings, each based on a cluster of trial sites. These were held during the period prior to harvest to discuss progress with the trials up to that date. In one instance, farmers came and went during the meeting but an attendance of about fifteen was estimated, including some half-dozen women. This was more rewarding than larger and more formal events. All the farmers came from the adjacent area and had seen the trials during the season (but not all had trials on their land) and some lively debate ensued. Encouraged by feedback from the farmers, who liked the new sorghum and maize cultivars introduced in the experiment, the trials assistant was able to persuade the local cooperative marketing depot that they should stock these cultivars. Attempts to encourage this from the top through the marketing agencies had previously been unsuccessful (Edwards, 1987b:12).

Edwards reports that another useful result from this meeting is that information on how cassava was being managed in the area was obtained for commodity specialists at the experiment station. When the question was raised, there was again a lively debate, and Edwards makes a comment which provides a final indication of the value of farmers' groups. The very liveliness of the meeting (which would not have occurred where local headmen were present in a formal capacity) gave 'greater confidence in the information ... about what was happening to cassava'. Farmers were not just repeating the standard explanation of what they thought the researchers wanted to hear.

3.4   Experience with group techniques in Colombia

JACQUELINE A ASHBY, CARLOS A QUIROS, YOLANDA M RIVERS

Group techniques for crop variety selection: snap beans

In a bean and cassava programme conducted in 1985 in Pescador, Colombia (as described in section 3.2), after the farmers had completed the first pre-screening stage of the on-farm trials, group methods were developed to help in variety selection. For example, a group meeting of farmers and researchers was held to discuss snapbeans. Because this type of bean was new in the area, these farmers were by definition experimenters; so the group was an 'innovator workshop'.

Ten farmers experienced in snap-bean production met for two hours with the project agronomist and anthropologist to discuss experience with the crop and to give their views on strategies for testing new snap-bean varieties on-farm. There was a lively discussion of different farmers' experiments and experience with planting distances, fertilization, rotations, disease control and marketing problems. The research staff used a checklist of topics to focus discussion and the group came up with recommendations for two types of on-farm varietal trial. A key preoccupation of farmers was the local scarcity of stakes for climbing snap-beans. They suggested one trial in which snap-bean varieties would be planted in rotation with tomatoes, utilizing the residual tomato fertilization and standing tomato stakes for support of the beans, which thus determined planting distances.

The second type of trial suggested by farmers entailed planting on a newly ploughed field, taken in from fallow. Fertilization levels at planting were suggested by the group and planting in double rows at distances which would facilitate staking, top dressing, spraying and harvesting was also determined by group discussion.

The group gave priority to the first type of trial – planting in rotation with tomatoes – as the form in which local farmers were most likely to plant snap-beans. A regional on-farm varietal trial incorporating these suggestions was planted in October 1986, to evaluate 15 climbing and 10 bush varieties of snap bean.

The group who helped plan this experiment later reconvened at the regional trial site and walked through the rows of beans examining plants and pods. The research staff asked farmers to show them what features of each variety they viewed as positive or negative and to indicate which varieties they considered should continue to be tested and which should be discontinued. Farmers' discussion rapidly focused on quality characteristics related to market acceptability of snap-beans. A list of these characteristics as explained by farmers is shown in Table 3.4. Farmers also identified two climbing varieties and two bush varieties which they viewed as outstanding in terms of the above criteria.

127

Table 3.4: Results of innovator workshop on snap beans
Farmers' criteria for selecting acceptable snap bean
varieties

---

1 Snap cleanly (without fibre) when pressed length-wise between finger and thumb; soft, non-fibrous texture (test is that the thumb nail should enter pod cleanly showing no fibre).
2 The bean should be cylindrical, not flat (plancha); long (approx 20–26 cm); and straight not curved.
3 Bean pod should be disease-free.
4 When snapped open the pod should be 'full' (llena) ie no tunnel of air appears inside between the incipient grain and the pod wall. This affects bean weight.
5 Deep green colour preferred (not pale green, or reddish, or purplish).
6 Yield (only if other quality requirements are satisfied).

It is noteworthy that other than the farmer who planted the trial, none of the farmers in the workshop had participated previously in this type of evaluation, or in farm-trials with the project. It appears that the group process, in which farmers interacted as much or more among themselves as with research staff, in discussing the pros and cons of the different varieties in the trial, was catalytic in motivating the farmers to undertake selections and to reject confidently a high proportion of the material included in the trial in front of the researchers. Moreover, involving these same farmers as a group in setting up the trial was important in the subsequent group dynamics. The farmers identified with a common objective – finding ways to introduce the snap-bean crop into their system – and by the time the evaluation was conducted, had already had the experience of taking part in the research process as a group.

After completing their varietal selections, the innovators' workshops continued: farmers were experimenting on their own farms with alternatives to staking with canes cut from local bamboo and wanted to set up trials to compare different planting distances and methods of support for climbing snap-beans. Also farmers wanted to compare different fertilization methods, a particular concern if they did not rotate snap-beans with tomatoes but planted newly cultivated fallow plots. As a result of the workshop farmers were creating initiatives for the formal research system to respond to by developing recommendations or a technology package for snap-bean cultivation, that would be the product of interaction with the group of experimental farmers.

Group meetings for bush bean and cassava evaluations

A second example of the use of participatory group techniques was conducted with expert bean farmers who had already taken part in on-farm

128

varietal trials. The objective of the group meeting was to synthesize the 'preference evaluations' and harvest evaluations earlier conducted with individual farmers and to ask farmers what they consider desirable criteria for selecting new varieties for future trials. On this occasion, men and women formed separate groups, discussed criteria which were listed on a large sheet of paper and then set priorities among criteria by voting on their importance. These are shown in Table 3.5 with priority criteria in italics.

Table 3.5:   Group evaluation: Farmers' characterization of preferred type of bean variety (priority criteria *in italics*)

Criteria given by men's working group

---

1 *High yielding*
2 Long pod with 6–7 grains (related to high yield)
3 Tall erect plant (not sprawling) appropriate for planting higher density
4 Adaptability to different soil fertility conditions, or fertilization
5 *Large grain size*
6 Deep red grain colour ('radical' type)
7 *Shorter season* (not longer than 85 days)
8 *Disease resistant* (1 or 2 sprayings adequate, not more)
9 Resistant to storage pests
10 Pod which does not split open in the field causing grain loss at harvest
11 Flavour
12 Soft-skinned when cooked
13 *Stability of yield* over at least 3 production seasons

Criteria given by women's working group

---

1 Quick cooking
2 *Grain swells* quickly, increasing total portion size when cooked
3 *Flavour* (sweet, not bitter)
4 Soft skin
5 *Resistant to storage pests*
6 Pod which is not difficult to open for threshing
7 *High yielding*
8 *Short season*

As can be seen from Table 3.5 the group produced specific criteria which for the most part included those identified from the content analysis of preference evaluations, shown in Table 3.3, but with a much lower investment of time on the part of research staff. In addition the prioritization of criteria provided some useful guidelines for the selection of varieties for inclusion in future trials.

A third example of participatory group evaluation was to identify objectives for further on-farm experimentation with cassava varieties. The group of local farmers identified as expert cassava growers in the local community met with the project agronomist and social scientists to review results of on-farm trials and to discuss directions for future experimentation. Trial results showed that the farmers' local variety out-yielded new lines at low and intermediate levels of fertilization; but that CIAT Hybrids and introduced varieties out-yielded the local variety at a high level of fertilization.

In the past, it has been assumed that farmers would not accept new cassava varieties if on-farm trials showed that they were out-yielded by a local variety, but the group indicated that this is not necessarily so. Farmers said that flexibility of harvest date would be an important factor to be considered over and above comparative yields, when determining acceptability of new varieties.

A first priority for evaluating the varieties further according to the group was to determine whether the new varieties could be harvested at earlier dates than possible with the local variety. In the local farming system, cassava ties up land and capital for a minimum of 14 to 17 months before it can be harvested. Farmers were interested in identifying which if any of the new cassava varieties could be harvested earlier than 14 months and whether these would spoil if left longer in the ground. New varieties with this potential would provide them with flexibility to respond to price fluctuations in the cassava market. If the new varieties had flexible harvest dates, then farmers might apply high fertilizer rates to obtain improved yields.

In further discussion, farmers were explicit that the cassava crop takes low priority in management in their system. Other crops which require intensive care are weeded, fumigated or harvested according to a specific calendar, whereas they only weed cassava when there is time in between other activities. Farmers' discussed their observations that timely weeding significantly increases yields, but only three farmers out of 15 were interested in an experiment to evaluate time of weeding. There emerged from the group discussion a self-defined 'recommendation domain' for which an experiment on weeding practices would be appropriate, but which received low priority from the group as a whole.

The group did agree however, that it would be important to re-evaluate the feasibility of timely weeding if the desired flexibility of harvest date could be identified in a variety requiring high fertilizer rates. Their traditional management strategy of low-input, land extensive, serial plantings of cassava to obtain varied harvest dates could then be called into question.

130

Review of experience in Colombia

The experience of the project in Colombia with farmer evaluations of varietal materials shows that this activity can provide breeding programmes with important information to streamline the selection of new varietal materials for specific farming systems. Although screening bean varieties with farmers did not have the objective of getting farmers subsequently to adopt them, follow-up observations showed that the small-grain bean varieties continued to be planted in farmers' fields, and were being disseminated from farmer to farmer, with women playing a significant role in sustaining this preference. The results of the group participation by farmers illustrates some of the advantages and pitfalls of the approach. Group participation in problem diagnosis has certain advantages for the efficient use of research staff time. A synthesis of farmers' common practices and alternative management strategies, which is the core information aimed for by informal surveys, can be rapidly achieved. The advantage of a participatory group diagnosis such as that carried out with the snap-bean innovators, as opposed to diagnosis conducted only by researchers, is that it is interactive. Researchers can test their interpretations of farmers' problems and even potential interventions to solve these problems, in a setting which stimulates farmers to discuss among themselves as much as to respond to researchers' questions. Consensus and dissent within a group are highly productive in highlighting farmers' management problems and constraints and different strategies for coping with them. The group provides a forum for prioritizing problems or needs from the farmers' point of view and is productive of conclusions about what new directions in technical innovation farmers themselves see as interesting.

The participatory group evaluations conducted in this project show that this approach can be usefully applied to identifying potential objectives for on-farm experimentation. Although group evaluations can provide researchers with ready access to farmers' ideas about desirable varietal characteristics or about combinations of technology components that are of general interest to farmers, the composition of the group is evidently critical in determining what priorities will be established by the group process. Criteria for defining group composition needs to be well thought-out for group participatory evaluation to be useful in orienting on-farm experimentation and establishing these criteria requires a careful knowledge of the farm community. The more homogeneous the group in terms of self-defined interests and perceived problems, the more effective the group process is likely to be.

For example, the bush bean evaluations carried out in this project identified farmers individually and in groups as 'expert' bean farmers. However, results of the preference evaluations and group discussions indicated that there are two distinct types of bean farmer in the community and that farmers define themselves and others in these terms: 'commercial bean growers' who plant larger areas and are primarily oriented to production for the market; 'farmers who grow beans mainly for consumption purposes'. The combination of these different types of farmer into a single group helps to explain why the varietal evaluations produced a mixed set of preferences, for large-grain varieties favoured in the market, on the one hand and for high-yielding, flavourful small grain varieties liked for their consumption qualities on the other hand. One conclusion that can be drawn from these results is the importance of consulting farmers about their perceived identification with different interest groups, when drawing up the criteria for the formation of groups for participatory evaluation of new technology.

The experience of this project indicates the practical value of giving farmers the opportunity to pre-screen a wide set of options among potential technological introductions. Group meetings showed what farmers wanted. This prompted the project to include a greater diversity of varieties and breeding lines in future trials.

3.5 Innovator workshops in Bangladesh

ZAINUL ABEDIN AND FAZLUL HAQUE

First experiences with farmer-led workshops

Though many scientists find it difficult to learn from the experience and knowledge of farmers (Chambers and Jiggins, 1986) many others recognize that there exists tremendous scope for such learning. But there are procedural and institutional issues and the question of social status cannot be neglected. Can the scientists sit in the classroom as trainees while an illiterate farmer is teaching? How do we identify an innovation? How do we learn the details of the innovations? What do we do with the new knowledge? (Brammer, 1980, 1982).

In 1981, one of us (ZA) was travelling with the Bangladesh Minister for Agriculture in the north-west of the country when a District Officer suggested a visit to an area where farmers were reportedly harvesting potatoes twice from the same planting and at the same time growing different inter-crops with the potatoes. A field visit to the Ranipukur area revealed that large areas had been planted to potatoes for double harvesting. Crops like cabbage, radish, wheat and chillies were being grown as inter-crops. The practice was unique and a decision was taken to show it to other farmers, extension workers and researchers, to whom it became known as the Ranipukur technology. A field day was organized and an innovative farmers' workshop was held at Ishurdi, Pabna in 1982.

Since this was the first known innovators' workshop in Bangladesh, there was no prior experience on which to base its organization. At that time, too, the term 'progressive farmer' was used instead of 'innovative farmer'. The objectives of the workshop were:

• to develop an awareness of the existence and use of progressive farmers in the quest for technology development;

132

- to exchange data and compare experiences on techniques of multiple cropping using potatoes as the base crop;
- to suggest how the existing arrangements could be improved;
- to see what lessons could be learned from progressive farmers for use by extension workers and other farmers and to identify areas for location-specific research.

The Thana Extension Officer was assigned the task of sitting down with the farmers and helping them to prepare brief papers on the technology they had developed. Four farmers were invited as resource persons and about 30–35 extension workers were selected as trainee-participants. Though many people feared that university-trained scientists would not actually sit as trainees and listen to farmers, in this case all the participants were fully motivated and took part with great enthusiasm. To them it was a big change.

This was a two-day programme. On the first day, about three hours were kept for an overview from the senior extension officers and a scientist. The remaining one and a half days were devoted to oral presentations by the farmer resource persons and also to group discussions and practical demonstrations. Informality was maintained all through so that the farmers were free of tension. At the end, the workshop came up with recommendations for adaptive research, extension and investigation by production economists (Abedin, 1982).

Mustard, wheat and watermelon

Following the success of the 1982 innovative farmers' workshop, it was decided by the Extension and Research section of the Bangladesh Agricultural Research Institute (BARI) that another workshop would be organized in 1983 on farmers' innovations in growing mustard, wheat and watermelon. The date had to be fixed for after a planning exercise by BARI's Programme Review Committee. It is worth noting that one outcome of this exercise was the suggestion that researchers should test a new variety of mustard *Sonali sharisha* (SS75) to see if it could be grown under zero tillage conditions like the traditional varieties. About a week later, the innovative farmers' workshop was held, again at Ishurdi and, during it, a farmer from Natore surprised everybody by saying that he had been growing *Sonali sharisha* successfully under zero tillage conditions for the last two years, but with one irrigation. A very good yield was obtained, but he did not apply any phosphate or potassium fertilizer because if the recommended rate of phosphate is applied to the preceding *Aman* rice crop, further application is not required in mustard and the response to potassium fertilizer is generally low.

In the same workshop, it was reported by the farmers that sprouting of watermelon seeds could be hastened by:

- burying the seeds in cow dung heaps;
- burying in earth near the *chula* (oven); or
- first soaking the seeds and then tucking the pack of seeds in the *lungi*

where it is in contact with the body. The body temperature keeps the seed temperature high causing sprouting during the winter months.

The farmer innovators also reported that wheat and watermelon could be grown together utilizing the land which remains fallow in the watermelon field.

During the same year, scientists were thinking about the possible performance of the same new mustard variety as an intercrop with pulses, a traditional practice in the pulse growing areas of Bangladesh. During a field trip, it was revealed that a few farmers about 10 kilometres from the research station at Ishurdi were already growing the variety as intercrop with lentils.

Bangladesh suffered a lot during 1984 due to flooding. Farmers had to transplant twice or three times. During a field trip in the Tangail/Jamalpur area, informal discussions were held with farmers trading rice seedlings on the highway. It was discovered that since there was still a risk of another flood, though it was September, the farmers were transplanting *Aman* paddy in the highland where mustard is usually grown after harvest of jute and keeping the traditional rice area on lower land free for mustard. This would allow them to grow both rice and mustard. Under normal flood conditions, growing rice on high land without irrigation would have been almost impossible.

These are only a few of the numerous examples of farmers' innovations. The point to emphasize here is that information about such innovations could be obtained by various means, but it is important that the effort should be made to obtain it. Many scientists and extension workers must pass through Narshingdi on the Dhaka-Sylhet highway, but how many have really observed that the country bean (*Dolichos lablab*) is being grown using chilli plants to bear the load of the bean vines instead of bamboo stakes? For such a practice, a chilli variety with a tall, strong stem is clearly required.

Two further workshops and influence on research

In 1984, the Graduate Training Institute of the Bangladesh Agricultural University (BAU) at Mymensingh volunteered to organize an innovative farmers' workshop. The extension service was asked to identify innovators, but was not requested to follow the practice adopted in previous workshops of preparing a written presentation for each innovation.

The workshop had the ambitious aim of gathering together sufficient knowledge 'for a possible change in the future curriculum of BAU with respect to rice cultivation . . .' (Hossain and Islam, 1985), but in the event, this was expecting too much. The organizers tried to tackle too many innovations at one shot and there were too many participants (about 100). Farmers were given only 10 minutes for presentation and discussion and in fact used much of that time to enquire from participating scientists how problems they were facing in rice cultivation might be solved.

A workshop with more limited objectives and more time was organized by the On-Farm Research Division in February 1985 and was held at Jessore. The main subject was an innovation in wheat relaying developed at Kushtia, but there were also discussions of other crops. One farmer innovator for each crop was identified by the extension workers and the workshop was arranged to last for two days, with the same format as used at Ishurdi (Haque, 1985). Seven farmers attended as resource persons and 32 researchers and extension workers were there as participants.

Most of these innovative farmers' workshops have produced recommendations which have been incorporated into formal research programmes at BARI. For example, after the workshop on the double harvesting of potatoes, on-station research programmes made studies of potato varieties suitable for double harvesting and have investigated different intercrops. They also investigated the performance of a cropping pattern involving potato, garlic, pointed gourd (*Potol*) and a green manure. At Ranipukur, farmers had developed a potato and pointed gourd pattern. Garlic and green manure were added by the researchers.

Following suggestions by farmers, scientists investigated one aspect of the planting of potatoes – the effect of direction of placement of the cut side on the growth and yield of potato. This was conducted on station. No difference was observed due to direction.

Following the workshop at Jessore, research programmes have pursued several questions concerning wheat relay crops. For example, researchers have worked out the optimum and maximum overlap period between wheat and the preceding rice crop. They have investigated the rate and time of application of fertilizer, have undertaken seed-rate trials and have explored possibilities in other parts of the country for relaying wheat with transplanted *Aman* rice, or with broadcast rice crops.

Conclusions

It may be clearly seen from this that innovative workshops have significantly influenced the research programmes of BARI. At the same time, however, experience has confirmed the contention of some scientists that farmer innovations must be fully understood before they can be used more widely. Thus it is important to appreciate that time and effort are required if one is to learn from farmers exactly what they have achieved. Two methods may be considered.

First, in person-to-person contact, the researcher seeks an appointment with the farmer so that he can discuss the details of the practice and its implications for other production and consumption activities at length. No interview schedule should be used. The information obtained is written up and copies are made for colleagues and for the farmer.

The second method of learning from farmers is via workshops of the kind discussed here. They have so far been held mainly at research stations, but could be organized in a rural school or community centre, with farmer innovators as resource persons. Workshops should be very informal with a 'moderator' rather than a chairperson. The number of participants should not usually exceed 30. Ample time should be given to

135

the farmers to explain their practices, and everything should be done to give them confidence in meeting people to whom they normally just listen. Successful workshops have always been carefully prepared in two respects:

- identification of topics to be discussed and farmers to take part. This can be done by researchers, extension workers, or the farmers themselves;
- preliminary documentation. Some preliminary information is collected to establish the relative importance of the innovations to be discussed and thus plan allocation of time in the meeting. Preparing the documentation may require the kind of person-to-person meeting referred to above.

During any workshop, a careful record of the proceedings should be made so that documentation of innovations and their use is available for future reference. The need for further research to develop the innovations should also be discussed and a set of practical recommendations should be drawn up for the research and extension programmes.

3.6 Farmer groups for technology development: experience in Botswana

D NORMAN, D BAKER, G HEINRICH, C JONAS, S MASKIARA, F WORMAN

The group development setting

Since 1982, the Agricultural Technology Improvement Project (ATIP) in Botswana has been conducting on-farm research with resource-poor farmers. The goals of the research have been to develop improved arable production technologies and low-cost research and extension methods. Our point of departure was the farming systems approach, with its commitment to a 'bottom-up' perspective. This approach has much in common with the various kinds of farmer participatory research recently described by Farrington and Martin (1987), including the 'farmer-back-to-farmer' model (Rhoades and Booth, 1982) and the 'farmer-first-and-last' model (Chambers and Ghildyal, 1985).

In Botswana, ATIP has addressed the farmer participation issue by working with groups of farmers that meet on a regular basis to discuss farming problems and on-farm trials. Prior to this project, farmer groups had not been used much in farming systems work. Individual farmers had often been included in the early descriptive/diagnostic stage of a project, but then researchers had reverted to procedures more typical of on-station experimental work. Too often, also, the link between research and extension had been pushed into the background. Groups can be important in keeping farmers in the foreground and in creating a spread or multiplier effect with relevant improved technologies.

136

The context of ATIP's work with farmer groups is a concern with the development of arable farming, stimulated by the Government of Botswana's interest in equity and employment creation. Despite rapid economic growth largely due to diamond and beef exports, Botswana is plagued by low and erratic levels of crop production. Throughout the arable parts of the country, rainfall averages only 450–500 mm per year. However, in 1987 the country was in the sixth year of a drought, during which farmers had produced less than 10 per cent of the national requirement for food grains, leading to a politically untenable reliance on food imports and food aid.

Agriculture in Botswana is centred on small, mixed livestock-crop farming systems. Cattle are the backbone of the farm economy, contributing milk for home consumption and cash through sales. Sorghum is the main crop, generally grown in mixtures with cowpeas and melons. More than 90 per cent of the area is planted to sorghum-dominated crop mixtures. Seed is broadcast and ploughed in using a mouldboard plough. The average area cultivated is around 5 ha. Oxen, tractor and donkey traction are used, with an emphasis on the first two. Only half the households control their own traction, but most households have access to traction through hiring or cooperative arrangements. Fertilizers, herbicides and pesticides are used by very few farmers. Average yields of sorghum are approximately 200 kg/ha and the returns to cropping labour have been less than $0.10 per hour during the drought.

During the first two seasons of research, ATIP attempted to address low arable productivity mainly through investigations of modified tillage and planting practices. After an initial emphasis on different planting methods, attention was increasingly concentrated on double ploughing, with spring ploughing followed by a combined plough-plant operation after at least one rainfall. Some promising results were obtained, but there was a bias toward richer and male-headed households because the changes in tillage practice required control of traction resources. Also, ploughing is the only arable activity which is dominated by men. By the third season, it became obvious that some steps would be needed to redress the imbalance and broaden the base of farmers involved.

During the years in which on-farm trials were conducted, ATIP implemented complementary research on household circumstances, on heterogeneity in traditional production practices and on the impact of support systems on farm systems performance. This research identified ways of building on traditional practices and revealed a complexity in household-farm interaction which required increased farmer involvement in the design of trials as well as the assessment of trial outcomes. The research also made us aware that groups are already pervasive in Botswana villages.

Group formation and administration

Based on the above observations and circumstances, farmer group activities were initiated by both the Mahalapye and Francistown on-farm research teams during the 1985–6 seasons. The objectives and procedures for group formation and administration were somewhat different in the two locations.

In the Mahalapye area, groups were formed in three villages in direct response to special circumstances and access problems of women and poorer households. In addition to facilitating trials management, the groups were developed in order to create an opportunity for on-going dialogue about problems and opportunities and the advantages and disadvantages of different interventions. Group formation in the Mahalapye area was viewed as an 'institutional experiment' in the following sense: a methodological goal was to assess relationships between group composition and the dynamics of farmer interaction in a technology development context.

All three groups – and all group members – continued to meet during the 1986–7 season. Over time, somewhat less emphasis was given to discussion of general problems, while more attention was given to discussing options for farming systems improvement. The groups continued to focus on interventions which were of particular relevance for women and for poorer households.

In the Francistown area, one farmer group was formed during the 1985–6 season in order to test double ploughing under farmer managed and implemented conditions and to get farmer evaluations of the system through the season. During the season, the wider potential of group testing activities became obvious.

As an outgrowth of the double ploughing farmer group, three groups were formed during the 1986–7 season. One rationale in forming the groups was to expand and supplement the research programme which had been based on researcher-managed work tightly focused on a few research topics. The specific group objectives were:

• to test a broad range of innovations under farmer management conditions;
• to involve farmers and extension agents directly in the technology development process; and
• to determine what types of innovations were most appealing to different types of farmers.

Experience of groups in two areas

Although the reasons for forming groups were somewhat different in the two locations where we worked, in both cases group formation was motivated by an interest in increasing farmer participation in the technology development, assessment and extension process. In each of three villages in the Mahalapye area, 10–20 farmers were recruited to participate in monthly meetings and to implement farmer managed trials. As part of the 'institutional experiment', different types of groups were formed in each of the three villages. Recruitment was done on a quota basis taking into account the desired household circumstances. In Makwate village, two groups were at first formed, one comprising females from poor households while the other was based on representatives from households involved in several past ATIP experiments. For logistical reasons, the groups were later combined to give one large heterogeneous group. In Shoshong village, the group was based on representatives from small conjugal units and both spouses were encouraged to attend meetings. In Makoro, the groups involved just females and most were from female-headed households. Most of the individuals attending the meetings in all three villages were female.

Each group elected a chairperson and set its own meeting date. A topical agenda was prepared for each meeting, comparable to a simple checklist used for an exploratory survey. At the beginning of each meeting the farmers reported individually on their problems and trials. Each farmer had one or more trials, which served as a focal point for group participation. This was particularly important for farmers who otherwise did not feel like talking about their farming problems.

Starting mid-season, a series of field visits were made in order to stimulate discussion. At the end of the season, a formal assessment survey was administered on both the trials and the group process.

In the Francistown area village meetings were held to form groups in each of three villages in early spring 1986. At the meetings, prior trial results were discussed and the activities planned for the following season were introduced. Interested farmers were asked to attend a special meeting at which the full range of technology options available for testing were described. The farmers selected the options they wanted to test. Participating farmers were provided with the equipment (if needed), seeds and fertilizer required for the experiment. Essentially all the experiments involved a simple comparison of the modified practice (or crop/variety) versus the traditional practice.

Monthly meetings were held with the farmers and extension agent in each village in order to discuss progress, problems and farmers' observations. A baseline survey and a mid-season assessment survey were administered to quantify farmers' reactions and problems. For each trial the dates of all field operations were recorded and grain yields were weighed by ATIP staff. Field days were held in which selected farmers presented their trials and results.

Typology of groups in Botswana

After two seasons of formal group activities, ATIP is firmly committed to the use of groups in order to facilitate participation in the context of farming systems research. While the implementation procedures and evaluation of group formats are still evolving, we have started synthesizing our thinking about groups, farmer participation and their merit in the Botswana setting.

A tentative typology is shown in Table 3.6. This distinguishes between design groups, focused-testing groups and options-testing groups. Although not originally intended as such, the groups in a sense constitute a continuum in farming systems terms and with reference to farmer participation.

Table 3.6: Typology of Farmer Groups

	Design groups	Focused-testing groups	Options testing groups
Objectives	Farmer involvement in technology design	Discuss farmers' own problems. Measure economic benefits. Farmer assessment	Increased farmer and extension involvement. Large scale assessment
Number of trial types	1–3	4–6	10–12
Trial			
proposal	Researcher	Researcher	Researcher
selection	Researcher	Researcher/farmer	Farmer
management	Researcher	Farmer	Farmer
implementation	Researcher/farmer	Farmer	Farmer
Quantitative measurement*	Most	Intermediate	Least
Assessment			
researcher	Most	Intermediate	Least
farmer	Least	Intermediate	Most
Group			
size	2–3 farmers	10–15 farmers	25–40 farmers
nature	Homogeneous	Homogeneous	Heterogeneous
selection	Technical situation appropriate for design work	Socio-economic situation for targeted technology	Volunteers from village meeting
Frequency of meeting	2–3 times a season	Monthly in season	Monthly in season

* Relative to other types of groups.

Following efforts initiated at Mahalapye, the distinguishing characteristics of the design and focused-testing groups are the relative homogeneity
of circumstances among group members and a concentration on a relatively
small range of interventions. The main distinction between the design and
focused-testing groups is the greater role of researchers relative to farmers
in determining the agenda of the design groups and in assessing outcomes.
Farmers are participants in the process, but primarily in the role of advisors
and assistants. Researchers are the primary client of the design groups in
the sense that the main objective is to develop knowledge about the
contributions of components to modified production systems. Because
farmer assessment plays a somewhat smaller role, it is not necessary for the
design groups to meet on a regular basis.

The focused-testing groups primarily serve as a vehicle for organizing
and assessing farmer implemented trials. An important feature is the
opportunity for farmers facing similar circumstances to discuss and assess
the relevance of a limited number of options for improving their farm
productivity. While researchers make a priori assessments of the relevance
of proposed technologies with respect to technical feasibility and con-
sistency with resource constraints, the farmer implemented trials and
associated discussions are needed to assess the economic viability (under
farmers' management) and social acceptability of options. The focused-
testing format is particularly appropriate for screening technologies which
are outside farmers' normal frames of reference. The discussions in the
focused-testing groups also provide an opportunity for farmers to identify
additional options not considered by the researchers.

A major strength of the focused-testing groups is also a weakness; the
researchers try to target technologies to a relatively homogeneous group of
farmers. This can create problems in that farmers other than those
identified by the researchers might be interested in a technological option.
Also, the small groups do create pressure on farmers to implement trials,
resulting in a distorted picture of farmers' independent responses to an
option.

The options-testing groups therefore represent an important step in the
technology assessment process in which a wide range of options are
presented to a large number of volunteer farmers. This enables as
assessment of farmers' reactions to a proposal to try a option, as well as to
the option itself. With less pressure to implement, a better assessment can
be made of the social acceptability of an innovation. With larger
numbers of participants, greater emphasis can be given to farmer assessment. The inclusion of local extension agents enables them to become
familiar with new technologies before promotion through the extension
service.

In a conceptual continuum of groups, yet another type of group has
been identified – dissemination and monitoring groups. These are not
included in the proposed typology because they are most accurately
viewed as extension groups, not technology development groups. Dissemination groups differ from the options-testing groups in three
respects:

- only a limited range of the most promising options can realistically be promoted;
- the groups are organized and managed by the extension service – village extension agents with the support of subject matter specialists and the local farming systems team; and
- the emphasis is on facilitating exposure to new technologies rather than assessment of potential options.

Advantages of farmer groups

Farmer groups have a number of advantages. The main ones as far as ATIP is concerned are highlighted in the following discussion.

Improvement of dialogue The group format provides a forum for improving dialogue with and among farmers. Unlike the more common approach where two or three researchers talk to one farmer at a time, the ratio in group meetings is reversed with a larger number of farmers in relation to researchers. This can completely change the dynamics of interaction. Regular group meetings help provide solidarity for the group, create familiarity between the group members and researchers and provide unique insights about farmers' priorities and perceptions.

The group format also provides an efficient way of ascertaining con-sensus opinions about the relevance of technologies being tested. For example, a major constraint in Botswana is erratic seedling emergence due to poor soil moisture and the lack of control over seed depth placement. Several solutions have been examined, including double ploughing and the introduction of various hand and traction-drawn planters. In one village in the Francistown area, where most farmers plough with their own animals, a consensus quickly developed in favour of double ploughing. In another village, however, many farmers said they could not easily double plough because they had to hire traction. In that village, most group members expressed interest in a hand rotary injection planter. The farmers said they could hand plant when there were good soil moisture conditions, regardless of when their ploughing was done. In this example, reactions in the two groups helped the researchers to more quickly identify why and where different solutions were required to what seemingly was the same problem.

Improved efficiency of research resources A continuing issue for farming systems practitioners is the need to economize on resources in terms of time and logistical costs. The group format provides a way to economize on the use of time since trial designs can be proposed and discussed in group meetings. Moreover, group meetings allow farmers to consult with each other about trial objectives and implementation procedures, thereby increasing implementation rates and reducing implementation errors. After trials are implemented, the time required for farmer feedback is saved by relying on group discussions.

With reference to logistical issues, inputs can be distributed to the

142

farmers at group meetings. Later in the season, schedules for data collection can be more effectively coordinated through joint discussions with researchers, farmers and enumerators. This is particularly helpful when different trials sites are planted over a period of several weeks to even two or three months – as in Botswana.

Facilitating farmer field days The farmer groups and associated trials provide an admirable format for farmer field days. In the field days, group members are encouraged to explain what they did in trials and why and what results they observed. The field days seem to engender a competitive spirit and to create momentum for the interventions which look favourable to some farmers. Although it is not necessary to have farmer groups in order to hold field days, ATIP researchers have observed that in field days dominated by representatives from groups, there is more discussion and greater momentum is achieved.

Potential for improving linkages To bring about agricultural development there need to be good linkages among farmers, researchers and extension agents. Unfortunately, in Botswana, as in so many countries, these linkages are not as strong as would be desirable. The group format provides an excellent opportunity for bringing together on-station researchers, farming systems workers, extenstion staff and farmers.

One of the main advantages of a group format is that researchers and extension officers outside the farming systems group, who are faced with limited amounts of time and resources, can address a number of farmers simultaneously. For example, this last year, groundnut researchers from the main research station were invited to discuss the value of fungicide seed treatment. As a result, a number of farmers tried a simple seed treatment trial and were quite impressed with the results. On the other side, the on-station researchers developed a greater appreciation for the farmers' current practice of planting groundnuts at very low populations.

The progress made in building linkages with extension agents has been one of the most obvious benefits of the farmer groups. By participating in farmer groups, extension agents collaborate in the development and assessment of technologies. Therefore, when technologies are ready for dissemination, the extension agents already understand any advantages and disadvantages and are in a better position to properly present recommendations to new sets of farmers.

Disadvantages of farmer groups

While there are clear advantages to well-functioning groups, not all groups do function well. In fact, groups of different types are pervasive in Botswana villages and the vast majority have severe problems which limit their effectiveness. This section reviews the major problems ATIP has thus far encountered in managing technology development and assessment oriented groups.

Inequality of dialogue and treatment In any group situation, not everyone will speak up. Since dialogue is extremely important, this can become a key group management problem. The larger and more heterogeneous the group, the less likely is it that all members will regularly participate in group discussions. One approach ATIP has used to facilitate participation is to have a portion of each meeting during which each farmer is asked to report on her or his own farming circumstances (such as any ploughing done). Even with such steps, however, there tend to be a few more articulate and aggressive group members who tend to dominate most discussions.

Another problem is a tendency to visit some farm sites more frequently than others. This can cause some jealousy. There is no easy solution for this where research resources are limited and not all trial sites are of comparable value in evaluating a proposed change in production practices. One potential solution, at least in the focused-testing format, is to distribute the hosting of trials fairly among group members.

The opportunity cost of meeting time Farmers subjectively evaluate the benefits from the time spent in group meetings relative to other activities. During busy parts of the year, the competing demands for farmers' time can lead to attendance problems. The ATIP on-farm research teams have tried several complementary approaches for maintaining farmer interest including: reducing the frequency of meetings during particularly busy periods and during winter (non-cropping season); providing transport for farmers living far from the meeting site; bringing in outside speakers; having refreshments at some meetings; arranging field visits or other outings of interest to the group members; scheduling meetings on days when drought relief food is distributed or when farmers traditionally do not work.

Group continuity and farmer replacement In ATIP's experience, nearly all farmers have wanted to continue participating in the groups. This has created an unexpected dilemma, particularly in the focused-testing format: should old group members be forced to drop out in favour of new farmers after two or three seasons? The main reasons for replacement are:

- the views and attitudes of old group members might become atypical due to ongoing interaction with researchers; and
- it is desirable to include as many farmers as possible in the technology development process.

On the other hand, it is difficult to exclude active and interested group participants. One of the main advantages of the options-testing format is that there is a less formal group structure, facilitating replacement on an annual basis. However, there can be a tendency for a gradually expanding membership which, in itself, can pose a problem in terms of the required research resources.

Farmer–researcher interaction Even simple trials have implementation requirements that force researchers to give some guidelines to farmers. If

144

meetings are dominated by researchers' presentations, farmers may adopt a passive role, and not shift easily back to a collegiate mode of interaction.

Outstanding issues

Some issues affecting farmer participation remain unresolved. The experiences of the two ATIP on-farm teams have often differed, but we agree that each issue affects the nature of farmer participation and needs to be addressed by farming systems teams contemplating the use of farmer groups.

Group size It is important to ascertain whether larger groups result in a lower quality of dialogue compared to smaller groups. If the quality of farmer dialogue is somewhat reduced in larger groups, is this a reasonable trade-off in order to enable more farmers to participate in farming systems activities? As reflected in the focused-testing versus the options-testing formats, the appropriate size of the group depends largely on the group objectives.

Researcher initiative Some degree of researcher initiative is inevitable in farming systems work since researchers often have information about options that fall outside the current scope of farmers' experience. However, the more researchers assume the initiative in group actitivies, the less collegiate researcher–farmer relationships become. Two issues need to be considered with reference to the degree of researcher versus farmer initiative: should researchers try to target options to particular farmer circumstances, as is implicit in the FSR recommendation domain concept? Do farmers have enough information about the potential options to assess a priori which should be tried?

The options-testing group format represents an attempt to shift the initiative from the researchers to the farmers by offering many options to interested farmers. In contrast, the focused-testing format is based on the assumption that greater targeting and researcher initiative is required when introducing (at least some types of) new options. Which format is more appropriate depends on the diversity of farmer circumstances, the type of options to be considered, and the objectives of group work.

Frequency of meetings Meetings can take up much time but, over time, fewer and fewer new insights are gained from discussions of general farming problems and there is less that needs to be discussed after an initial trial implementation period. The issues are whether it is necessary to meet regularly and how often to do so. The frequency of meetings obviously depends on farmer interest, but also on the success of researchers in arranging supplementary activities like field visits or presentations on specific topics of interest to farmers – such as a demonstration of how to spray sorghum for aphid control.

145

Conclusions

ATIP experiences indicate that groups can be effective in increasing and improving farmer participation in the technology development process. Groups keep farmers in the foreground, provide a means of using social dynamics constructively and create a multiplier effect with reference to the spread of relevant improved technologies. There are many other benefits from farmer groups, including the increased efficiency in use of research resources and improved linkages between researchers, extension agents and farmers.

The idea of farmer groups has struck a chord within the farming community. Farmers have almost universally expressed an interest in continuing in the farmer groups. Although some problems and issues remain, ATIP's experiences suggest that farmer groups provide a pragmatic tool for undertaking farming systems work which is complementary to informal and formal surveys and researcher managed trials.

The formation of farmer groups should be seriously considered by other FS teams concerned with the issue of farmer participation. Several group formats have been discussed on the basis of experiences in Botswana. While these group formats have worked well in Botswana, the structuring and management of groups obviously needs to be adapted to different social and agricultural settings.

3.7   Participatory technology validation in highland communities of Peru

MARIA E FERNANDEZ AND HUGO SALVATIERRA

Labour bottlenecks and livestock

This paper discusses a multidisciplinary experience in three high-altitude peasant communities of the Mantaro Valley in central Peru, where participatory research has been done with men and women farmers over a period of three years. Adaptive trials are at the centre of the action. Which trials to carry out – whether on control of livestock parasites or on crops – has been a decision arising from a group participation exercise. Alternatives were suggested from farmers' experience as well as by researchers and were pre-screened by farmers' groups and researchers together.

The background to this research was the experience of the 1950s and 1960s when many 'improved' agricultural technologies developed by experiment stations were not adopted by small farmers. The common explanation then was that the small farmer was too traditional to accept new ways. He or she was thought to be complacent with the status quo, comfortable with a limited standard of living and averse to change.

Although most crop and animal research is still carried out within the confines of experiment stations, we have come to realize that 'traditionality' is not the main reason why the small farmer does not adopt improved technologies. It is becoming more and more evident that many of the new alternatives require capital, labour and ecological conditions to which the subsistence farmer does not have access. Taking these factors into consideration, the limitations of research methodologies in current use are being more widely appreciated and institutions such as the International Potato Centre and pilot projects within the Peruvian Ministry of Agriculture are attempting to carry out experiments under conditions more similar to those faced by small farmers.

The most important problem confronting the small farmer in the Mantaro Valley is the high labour intensity of most operations. People must be available to do essential tasks at the right time. In small-scale production systems, this factor is critical as all community members are subject to similar labour demands during the same periods of time when neither cash nor paid labour is readily available.

The high-altitude small farmer has evolved production strategies and organizational forms which help him to overcome some of these bottlenecks. One of these is the complementary management of crops and livestock in the system which permits greater independence from external inputs. Shortages of labour are overcome in part by an efficient distribution of responsibilities for animal and crop production activities related to the interaction of gender and age groups within the production unit, by allocation of labour tasks between the same groups and by shared labour arrangements with other community members.

In order to provide the small farmer with technologies which will improve production, it is necessary that these be designed to take into consideration the combination of ecological, technological and organizational factors which have commonly been the subject of crop and animal research. Arising from this are a series of questions, the answers to which we consider basic to appropriate research with small farmers:

- Who is the producer? Who is making the decisions in the production and sub-production units?
- What are the principal problems or limitations which the producer is attempting to overcome?
- How are the production tasks distributed among men, women and children?
- What experimental designs will serve at one and the same time to prove the effectiveness of a technology and to demonstrate its benefits and limitations to the producer?

If it is true that the agronomist, animal science specialist, veterinarian and extensionist have a wealth of knowledge concerning biological factors related to production, it is also undeniable that the small farmer has a wealth of knowledge concerning the management of ecological, technological and organizational factors related to production under specific conditions. In the Andes, where land, crop and animal management systems have been highly developed for at least four centuries, the small

147

farmer has access to a systematic and historic body of knowledge which may influence his production practices.

Participative research, then, is a means by which two bodies of knowledge can be brought together and can interact so that the solution of small-scale farming problems can take place over a shorter period of time than in conventional research and with greater confidence that the results will be adopted.

Participation in this sense not only means that small farmers play a practical role in research by planting trials on their land, but that they also discuss how those trials will be conducted both individually and in group sessions. In general terms, we sought farmer participation in *defining the problems* we would tackle, in *designing experiments* relevant to these problems, in *implementing* experiments and in *evaluating* results.

As essential background to this work, we attempted to understand in detail how responsibilities were distributed by gender and seniority and by this means to find answers to the first of the questions posed above. For example, we enquired about how people approached decisions such as when to sell an animal, or when to sell grain or potatoes, or whether to purchase inputs. We asked whether there was general discussion within the family group, or whether an individual took the decision alone. We also tried to distinguish between decision-making, technical specialization and responsibility for carrying out tasks. In particular, did women have access to all aspects of knowledge, especially concerning animal production? Could women enter into all aspects of agriculture?

Farmers' reactions to plans for experiments

When the implementation of on-farm experiments was planned with farmers during the first year of our work, we were surprised by some of the problems encountered. When we wanted to do crop trials, there were objections to leaving spaces between plots and to anything which added to labour requirements. When we proposed experiments on parasite control in animals, we found that no individual family herd was big enough to carry out a complete experiment. In addition, the farmers were unwilling to treat only selected members of the herd. When experiments on endoparasites were proposed, farmers were reluctant to sell animals to be slaughtered so that the appropriate laboratory work could be carried out.

The objection to leaving spaces between plots was that the farmers regard unplanted areas as a waste of utilizable land and of soil nutrients. Land planted with potatoes in the second year of the rotation cycle is fertilized with the intention of sustaining subsequent crops also. Not cultivating it therefore entails loss. In addition, land not cultivated encourages weed growth. If there is a need for spacing, this is only acceptable in grain crops which come at the end of the cycle.

Spaces between blocks and treatments also make work with the ox-team difficult. This means that the participating farmer must prepare the land with a pick, which requires more time and effort and more personnel. The use of short furrows requires reloading the *quipi* which the person doing

148

the planting uses to carry the seed, adding significantly to the time taken.

Planting is done by a team which must work in close coordination. The ox-drawn plough goes ahead opening the furrow, followed by the person who places the seed in the ground, and then by the one who distributes the fertilizer. If any of the three takes longer at his or her task than is usual, the work of the rest is disrupted. For this reason, the distribution of small amounts of seed of any variety requiring the sower to reload the *quipi* presents considerable problems for the overall work rhythm. A similar situation arises with fertilizer treatments, where distribution is done using hand measurements and variation in the quantities requires extra actions.

As a result of these problems and incorporating the suggestions of the farmers, we have made adjustments in the experimental designs which permit control and measurement while allowing the work of planting, harvest and animal management to be carried out normally. The following short case-studies illustrate some of the experimental designs worked out with the farmers for trials on crops and animal production.

Redesign of experiments in response to farmers' objections

In order to avoid the uncultivated spaces between blocks and treatments, an attempt was made to plant *tarhui* (an Andean legume) as a divider. At harvest, however, the ripening of the potatoes and tarhui did not coincide and the ox-team killed the tarhui when loosening the potatoes. During the second year, *mashua* (an Andean tuber often planted in association with potatoes) was substituted for the *tarhui*. The process consisted of planting an entire *quipi* of potatoes of the variety desired in the prepared furrows and then placing three or four tubers of mashua before reloading the *quipi* with another variety or treatment. The resulting distribution of tubers across the field would then be as shown in figure 3.1. This system avoids loss of time at planting and permits a simultaneous harvest. In subsequent experiments, it was found that five tubers of mashua is the minimum which guarantees that the treatments will not be mixed when the ox-team loosens the tubers at harvest. The only special care that must be taken is to note carefully the direction in which the planting team has worked (up one furrow and down the next) so that the identification of blocks and treatments is not confused.

A second way to avoid leaving land uncultivated is to plant the tubers for each treatment in a series of complete rows. If this is done, the number of rows must coincide with the quantity of tubers in a *quipi*. For potatoes, this implies a minimum area of 20 square metres and requires a larger quantity of seed for each treatment and block than for the previous approach. It is a method which can also be used for experiments with maize, peas, *quinua* and *tarhui*.

When experiments are done in plots of barley, wheat and oats, the local broadcast method of sowing has to be taken into account. The field is divided into contiguous strips marked only by a shallow furrow at the edge of each left by the ox-team. The strips are 3–4 metres wide and the sower

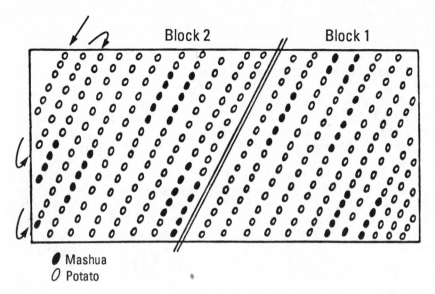

Block 2   Block 1

● Mashua
○ Potato

Figure 3.1: *The method of separating blocks and treatments for on-farm experiments with potatoes in the Mantaro Valley, Peru.*

limits his distribution to this width. Blocks and treatments are planted in complete strips and the evaluation is done in the centre of each to avoid any mixing at the edges.

To alleviate the problems mentioned in animal health experiments, several modifications were made in procedures. Family herds were used for each block or treatment. Care was taken to choose herds grazed in similar areas of the communal range during the same number of hours daily. Each herd, however, was kept in a different corral at night. Here, the entire herd (averaging 25 sheep) was given the appropriate treatment. In cases where slaughter became imperative, the chosen animals were replaced by younger ones of the same sex by mutual arrangement with the farmer.

These examples serve to show that farmer participatory research is not just a matter of group discussions and asking farmers to plant trial plots. Researchers must be willing to re-think their whole experimental procedure and the design of some experiments requires very careful planning. This is necessary in order to obtain quantitative results and comparisons with controls while at the same time fitting in with the farmer's own methods and allowing him or her to make an individual assessment. Like other researchers, we have found that farmers' evaluation criteria are often very different from ours. In addition to yield, potatoes are evaluated for their colour, for resistance to pests, frost and hail, for storing well and for taste and texture when cooked.

3.8 Two complementary approaches to farmer involvement: an experience from Indonesia

CAROL J PIERCE COLFER
with FAHMUDDIN AGUS, DAN GILL, M SUDJADI, GORO
UEHARA, AND M K WADE

The site and its problems

Sitiung, in West Sumatra, is a transmigration area of 100,000 ha composed of more than 12,000 resettled families interspersed with nearly as many of the indigenous Minangkabau (or Minang) inhabitants. The area was originally lowland humid tropical rain forest and much of it remains forested. The soil is extremely infertile with high levels of aluminium toxicity. Despite 2,500 mm annual rainfall, periodic dry spells and resulting water stress in plants add to agricultural risks.

The indigenous Minang farmers consider rubber and paddy rice to be their mainstays, but pursue many other economic activities. The transmigrants from overcrowded Java are initially given 1.25 ha each on which they grow rice, peanuts and soybeans with a wide variety of other crops in their home gardens.

Tropsoils, a USAID collaborative programme on soil management, began work in the area in 1983 with three Americans (an agronomist, a soil scientist and an anthropologist), five Indonesian soil scientists and five agricultural technicians. We arrived on site later than intended and rather than miss the rice planting season, we chose a village for our work and farmers to collaborate with, in a less-than-ideal manner.

We wanted to use Hildebrand's (1979b; 1981) *sondeo* method to choose a site for our work. However, the complexities of getting settled and dealing with a soil survey that was under way resulted in our doing only a very modified *sondeo*. On the basis of three days of intensive interviewing and interaction among ourselves and the villagers, we chose to work in Aur Jaya. This was the most recently settled area, and we reasoned that documenting the adjustments of the settlers from the very beginning as well as changes in the soil would be of value. The appearance of the location, littered with felled but burned logs, convinced us that there was a *need* in Aur Jaya which surpassed all other locations.

The ethnic mix of the community (which repeatedly proved to be significant in our work) was approximately 40 per cent Sundanese from West Java, 40 per cent East Javanese, and 20 per cent from the nearby Minang community whose lands were used for the settlement. In June 1983, at the close of official settlement, the total population numbered 1,466. The proximity of the Minangkabau settlers to their home area and their access to significant amounts of land there initially resulted in a pattern of dual residence. However, after the government subsidy was terminated in 1985, most of these people abandoned their Aur Jaya homes completely and resumed permanent residence at their original homes.

Recognizing the equity-related pitfalls of choosing cooperator farmers via existing community leaders (all male, generally better educated and wealthier than average), we tried hard to choose farmers ourselves on an individual basis. However, in the end the planting season was upon us and we were forced to turn to the leaders. We reasoned at the time that it was better to choose unfairly the first year and get started than to wait a full season.

Having identified four geographical sub-communities in Aur Jaya, we approached the four leaders who corresponded to those communities and asked each of them to suggest five farmers who might want to participate with us in our experiments. In this way, the overall ethnic composition of the community was replicated among our cooperator farmers, but the problems we anticipated when we regretfully asked local leaders for suggestions did occur. Two leaders chose themselves or close family members and then did not have the time needed to devote to agriculture. They suggested friends and clients in such a way as to coopt us into their patron-client relationships, thereby reinforcing their own political standing. They also chose male heads of household and while there were few woman-headed households at that time, we did wish to involve women in a meaningful way.

We next called meetings of each group of five cooperator farmers. Their wives were invited, indeed urged, to come, but did not. In these meetings we explained the reason we were working there, including our goals of developing agricultural technology that was usable by and of interest to farming households and of learning from the farmers' knowledge and experience what could be included in our research. We also explained our philosophy that whatever we developed should be affordable by the farmers themselves. All of the 20 farmers contacted in this way chose to work with us.

Starting the experiment

Although we had an experimental plan in mind which included four soil amendment treatments in upland fields, we wanted to get farmer input at a large meeting of all 20 farmers. At the meeting we carefully explained that our plan was only a suggestion and that we wanted to put our scientific knowledge together with their experience and concerns. Moreover, the experiment could be arranged to cover a whole field, or just a plot measuring 10m × 20m. We also proposed a sequence of crops: rice, later intercropped with cassava and then after the rice had been harvested, an edible legume planted between the rows of growing cassava. After that, there was the possibility of one final crop.

We were surprised and delighted in the meeting at how readily the farmers came up with suggestions. They preferred the 10m × 20m plots rather than use of their whole fields and suggested variants in the cropping programme. We had proposed to end the cropping sequence with a green manure on half the plot in order to compare subsequent yields with the other half when that was used for vegetables. One farmer suggested that instead of an inedible green manure, we should plant mucuna bean which the East Javanese eat. This suggestion was finally adopted by the group and was a definite improvement on our plan.

We researchers had done some soul-searching about the provision of inputs and finally decided to provide fertilizers, lime and pesticide, because of our uncertainty about this new environment. We were also unwilling to subject farmers at the subsistence margin to any unnecessary risk. There was also some discussion of the logs which littered the farmers' rice fields. Analysing plots in this condition would be a statistical nightmare, but paying to have them removed seemed a bad precedent. In the end we compromised. We bought a chain-saw, hired workers to cut the logs into manageable lengths and then three of us helped the farmers to roll them off the fields. This helped in establishing rapport with the farmers and in getting to know them and their concerns.

The cropping pattern described earlier was imposed on four soil fertility treatments:

- no additions;
- government-supplied urea and TSP at 100 kg/ha each;
- rock phosphate at 800 kg/ha plus urea and TSP; and
- lime at 2.5 tons/ha plus urea and TSP.

The farmers agreed to hoe in the rock phosphate and the lime; the remaining plots could be hoed or not at the farmers' discretion.

We worked with the farmers daily during this time, laying out plots, delivering lime and fertilizers and helping to plant. During this process, three additional changes were made in the experiment, two of them connected with hoeing.

The first began with some griping from a normally hard-working Sundanese farmer who said that it was too hard to hoe the lime and phosphate plots as we had agreed. Soon other farmers were complaining and we began to think of the oft-repeated stories of farmer laziness. However, being out in the fields daily, we soon observed that the plots were a tangled mass of roots just below the surface. We saw that the Sundanese farmer's hoe could hardly penetrate them and other farmers had even broken their hoes. In view of this, we altered the experiment to specify incorporation of the fertilizers by light hoeing.

By this time, an important ethnic difference in land management strategies had emerged. The Javanese and the Sundanese were avid hoers, feeling that not to hoe was symbolic of a general lack of diligence. The Minang, on the other hand, used to comparatively extensive land holdings, never hoed unirrigated rice. They simply planted with a dibble stick after burning. Moreover, one of the Minang farmers was old, alone and weak. It became obvious that he was not going to hoe and two other Minang farmers seemed also to be opting out. This led to our second modification of the experiment, because after considerable discussion, we decided to incorporate this reluctance into the research design. We would compare three farms on which the phosphate and lime were not hoed in with the other farms where light hoeing had been completed.

We soon began to see advantages in participatory research. The hoeing variable, which the farmers had been concerned about from the first, turned out to be more important for crop yields than any of the original, researcher-designed fertility treatments. It seemed that cultivation could, to some extent, substitute for fertilizer inputs and further research was done to check this (Wade et al, 1985).

The third change in the experiment was the incorporation of corn, to be intercropped with rice. This was suggested informally by the farmers during rice planting. We agreed to the idea and explained the planting procedure at a meeting, but much of the corn was later planted by women none of whom had been at the meeting, so there were many inaccurately planted plots. Moreover, delay in planting meant that it could not compete with the rice and after being attacked by rats and mice, did not produce any significant yield.

Meanwhile, the farmers themselves were experimenting with various crops. Chillies and peanuts were growing well in home gardens and getting a good price. We were accosted with suggestions about incorporating peanuts and/or chillies into our cropping sequence (only later did we realize that the Javanese were growing peanuts, and the Sundanese chillies!). We agreed to this, but while the peanuts grew well enough to get reasonable yields, the chillies were another disaster. No one in the team and few of the farmers had enough experience with chillies. Our modest fertility programme – less than farmers had applied to their home gardens – was apparently inadequate and we did not know how to control diseases which proved very destructive.

We had planned a three-crop cycle with the final crop being a split plot of mucuna bean and mixed vegetables, but farmers failed to complete the activities required for the second crop in time to include a third. We also learned that farmers regarded two crops as the norm. Longer-term residents regarded a third crop as very risky. Despite having agreed to three crops at the beginning, the farmers were now letting us know by dragging their feet what their real preferences were. We also realized that growing vegetables as a third crop in the rice field was not a particularly good idea because the farmers' household gardens already produced such crops on an adequate scale for family consumption.

Research directions and a parallel study

Although we achieved good interaction with male farmers, lack of involvement with women was a drawback. Part of the problem was that most of our team members were young Indonesian men who were a bit shy with women. Even though women's status in Indonesia is comparatively high and their involvement in agriculture is accepted, unfortunately few women participated in our project. Our concern about this led to some activities focused on home gardens. Though women are very much involved in the fields as well, they do most of the work in home gardens, which are easily accessible to kitchens and lend themselves to management by women. Moreover, a greater proportion of both cash income and subsistence

154

production than we had originally thought comes from home gardens (Colfer et al, 1985; Chapman, 1984). In addition to the gardens, other research areas that our experience encouraged were tillage experiments (Fahmuddin Agus et al, 1985; Makarim and Cassel, 1985), and the use of green manures as soil amendments (Evensen et al, 1985; Gill et al, 1985). The work with farmers was not wholly responsible for what Tropsoils has done on gardens, tillage and green manures, but knowledge of farmers' conditions in Aur Jaya was a significant influence on the direction of our research.

One further new direction stemming from the Aur Jaya experience concerned the Minang farmers. By late 1984, their continued participation was in question. Most had returned to their home village. Our experience of working with them for one full year, however, had convinced us that there were some significant differences between their farming system and that of the transmigrants and it seemed very probable that we could learn something from farmers who had dealt with this particular environment throughout their lives.

One feature of their system that seemed a potentially important research opportunity was their dependence on tree crops. They grew rubber as well as paddy rice and there was general agreement that tree crops were better adapted to these fragile and infertile tropical soils than were annuals. However, none of our agricultural scientists knew about tree crops with saleable products.

By 1986, I was more than ever aware of the extreme riskiness and lack of profitability of annual crops in that environment, as well as the reluctance of agricultural scientists in general to take on tree crops in a serious fashion. Not only did trees reduce soil erosion and contribute to fertility via leaf litter, but the lower labour requirements and the greater income per person-hour of work seemed good arguments for trying to develop agricultural or agroforestry systems that included trees. The frequent occurrence of coconut, jackfruit, papaya and papaya leaves in people's diets also make trees a legitimate nutritional/subsistence concern (compare Passerini, 1986).

Recognition of such factors ultimately persuaded me to undertake a nine-month anthropological study in Pulai, a Minangkabau village not far from the transmigration site project. We were interested mainly in Minang interaction with their soil. I therefore sought out the situations in which Pulai people use their soil in some way – in the rubber orchards, the paddy rice fields and in their home gardens – retaining the usual anthropological conviction that problems can only be understood if they are considered holistically, as part of a complex of interacting causes and effects (see Vayda et al 1980).

Participant observation was my basic method and generally as the research progressed, I designed and undertook more focused studies on specific topics that emerged as important. I began with a year-long observational time allocation study designed to learn how people used their time in this farming system compared with the transmigrants at Aur Jaya.

We were particularly keen to document the involvement of Minang women in agriculture. Minang women are generally conceived to be primarily responsible for rice cultivation. This is partially related to the matrilineal inheritance of land. They also harvest and sell most of the produce from fruit trees. Women's daily involvement in all the stages of rice production presents an opportunity to work with them more easily than it was to work with the transmigrants who tended to consider agriculture as 'men's work' (despite women's active involvement).

Another study focused on indigenous knowledge of soils. By interviewing everyone I could corner and using ethnoscientific methods, I gained a reasonable understanding of Minang soil and land classification. Members of the research team took soil samples in the various soil types identified by the Minang, recorded the various crops growing in different kinds of land and soil, measured fields and estimated numbers of trees per hectare with indigenous planting patterns. The team also used a crop cutting technique for measuring farmers' rice yields. Then our final quantified effort was two surveys of all Pulai households: an ownership survey and an income survey.

We found evidence that although production levels from tree crops are low, given a very minimal management strategy people consistently get some income with very little effort. Some trees, such as rubber and coconut, serve as a savings account. Others, such as rambutan or coffee, provide a reliable if small seasonal income. Rubber and coffee can grow within the forest cover, maintaining a habitat for wildlife and for gathered foods like ferns and bamboo shoots while preventing the erosion that occurs on cleared sites. Apart from rice and fruit, fully 25 per cent of the village's income comes from paid work in very small-scale logging or in tapping forest rubber. Another 15 per cent comes from the villagers' own rubber and coffee 'orchards' growing within the forest. Thus the dependence of the Minang on the forest is clear.

Methodological comparisons and conclusions

Throughout the study of the Minang community and interspersed between the more focused surveys, I continued with participant observation, attending any community functions I could, following people around as they tapped rubber, weeded paddy fields, planted upland fields, processed coffee and so on and after every experience, I wrote field notes: what had I seen, who had been there, what was the work like, what did people talk about, what assumptions and values were implied?

The work on Minang agroforestry and the on-farm experiments on transmigrants' land at Aur Jaya represent two different methods for involving farmers in agricultural research. Both were of value and we deliberately chose to run them in parallel because they complemented one another, not only in terms of the information they provided, but also in relation to the complementary personalities and interests in our research term. Some team members were great at collaborative, on-farm research experiments. Others never really got into the on-farm work but were

156

intrigued by the different methods I used with the Minang and were keen to provide the technical, soils-related support.

Where collaborative on-farm experiments are possible, they seem to provide more direct, usable experience for scientists, with more immediate feedback on the topic that is most important to them. But without a resident person responsible for farmer involvement, there is a danger that such research can deteriorate into the traditional, undirectional communication from scientist to farmer. Or, since it fits awkwardly with traditional scientific practice (and prestige), it may be quickly shunted off to the most junior person around (who can also be forced to live 'in the field').

Effective involvement of farmers in agricultural research requires a real commitment on the part of the project personnel and administration. Someone must be identified as responsible for ensuring that it occurs, or it will simply fall away – because it is much easier for agricultural scientists not to do it, even if they think it ought to be done.

We have already mentioned some of the ways in which Tropsoils research benefited from the on-farm experiments, in terms of stimulus regarding tillage experiments and work on green manures. The study concerning Minang 'agroforestry' could provide stimulus of a different kind. Tropsoils has almost exclusively emphasized annual field crops, partly because that has been the subject-area on which our researchers have most experience. Yet such crops – compared to tree crops – require vast amounts of time, fertilizer and human labour. They are subject to many local pests. They are of little use in erosion control. They are very vulnerable to irregularities in the rainfall pattern. This study allowed us to document the differences between the farming system practiced by the Minang and that of the transmigrants. That was an important first step in persuading others to conduct work on non-annual and non-field crops. On-farm, collaborative research with Minang farmers would provide additional access to their knowledge of local conditions and allow scientists to begin working on components of a mixed food-crop and tree-crop system. But such work has yet to begin. Agroforestry systems, like working with farmers, are inconvenient to deal with. It remains to be seen what will transpire.

3.9   Final reflections about on-farm research methods

IDS WORKSHOP[1]

Complementary methods

Although this part of the book has again emphasized farmer-first approaches, with farmer participation and farmers' contributions to research, no author has argued that *all* agricultural research should be conducted with farmer participation. Some research has to be undertaken by scientists working at experiment stations or similar institutions. For example, in section 3.6, David Norman and his colleagues stress the need to improve linkages between experiment stations and farming systems research involving farmers' groups, to ensure that experiments done in each context produce results of mutual value. On-farm research and on-station research are seen as complementary and not in competition.

The complementary roles of different types of research are nicely underlined by Galt (1987) in reviewing practices in Nepal. Farming systems research (FSR) has had successes in that country, but that does not mean Nepal should 'integrate' all its research around an FSR framework, even if it could afford to. There are advantages in diversity. Regional research stations have a role to play in carrying out trials in diverse agro-ecological zones, just as on-farm research and group treks have their value.

In any case, there is no one method of doing FSR that can be held up as a model, so 'different approaches should be tolerated and encouraged ... In fact, if (practitioners of) each different approach are willing to listen and learn from each other, it is better to have different institutions and approaches' rather than to standardize. Those doing FSR in one or other of its forms will feed results from on-farm trials and group treks back to the station-based commodity programmes. The key to effective research is then 'willingness and ability to adapt that which is useful and relevant from one to the other' (Galt, 1987). Ensuring that these linkages are made between 'formal' and participatory methods raises institutional and policy issues to be addressed in Part 4.

There are also 'complementarities' between different participatory methods. No single method (such as an innovator workshop, creating a systems diagram, or an on-farm experiment, etc) need stand alone, but instead can be used with others. For any problem the best mix and sequence of methods will vary, depending on the resources available to researchers and farmers and local social and environmental conditions.

On-farm experiments: scope, methods and the systems approach

Questions raised repeatedly above concern the aims, methods and scope of on-farm experiments.

First, several authors have shown how experimental design can be modified to fit local practices. One example is the unusual planting pattern adopted in potato trials in the Andes (section 3.7). Another is how local views on the hoeing of rice-fields were incorporated into the experiment in Indonesia (section 3.8).

A second issue concerns statistics. Almost all the on-farm experiments described were designed to produce statistical data. Norman et al (1987) point out a need for on-farm numerical results that can be used by other agricultural institutions, but critical views have been expressed about statistical methods (eg, in section 3.1). Many would agree that the most important indicator in evaluating new technology is farmers' adoption.

Where other statistical results are needed the analytical rigour required differs according to whether the analysis is to help farmers or to help the experiment station. Techniques in experimental design and analysis which were *not* thought very relevant or helpful included standard randomized block designs, Latin squares, and factorial and multiple treatment structures with analysis of variance. In contrast some techniques found useful include scatter diagrams for presenting results from a number of farms and linear regression, for investigations into the stability of biological systems.

Third, questions have been raised about the form and scope of experiments – for example, asking whether experiments should be trials of one component in a farming system, or aimed at understanding the system as a whole. Most of the on-farm research described on previous pages has taken the form of trials on crop varieties, ploughing and planting techniques, animal health or fertilizer treatments. However, some practitioners of on-farm research argue that the notion of a 'comparative trial' between an improved crop or technique and a traditional one is questionable, because farmers rarely drop one practice in exchange for another. Rather, they fit new practices in alongside old ones, sometimes modifying both. Most of the trials described earlier take this into account to some extent, often by presenting farmers with a range of choices rather than just a comparison of new and old and expecting only some elements from that range to be adopted and then in an incremental way.

Fourth, conventional on-farm trials are only one of many possible points of entry. Conventional trials imply on-farm research where it is the business of the researcher to introduce some innovation that can be tested. If one thinks of the farming system as a whole, this is only one starting point. Another is to explore changing the relationship of components already within the system. Yet another is to find out what experiments farmers are already doing, and develop these. Or, more obviously, one can begin by working with and enabling farmers to identify problems for research. Thus, as shown by several contributors, there are several possible points of entry and alternative approaches and sequences apart from trials of new technologies. In many of these, identifying farmers' agendas (the theme of Part 2) and farmer participation (the theme of Part 3) are closely linked.

A learning process: lessons from the Philippines

This point is brought out well by Repulda et al (1987) in an account of work in six villages on the islands of Samar and Leyte in the Philippines. Cultivation systems there are complex and many different ecological niches are worked, ranging from shifting cultivation in rain forest to paddy rice on permanent bunded plots. Though average rainfall is high, there is a significant drought risk. Soils are erodible, with much steep sloping land and low fertility.

Repulda and his colleagues describe a farming systems project which over the last four years found several ways of working with farmers to tackle these difficult conditions. The project began with conventional, researcher-managed trials on farmers' fields, with new technologies, such as fertilizer, followed by trials of cropping patterns oriented to a FSR approach. However, for upland farmers, 'our cropping patterns were too expensive in fertilizer and pesticides; too demanding in time . . . and too demanding of the easily exhausted soil'. Even more significantly, 'mixing and matching of crop sequences must be highly flexible as changes occur not only in weather, but also in household food requirements, market opportunities, land quality, and weed species' (Repulda et al, 1987:4).

All these considerations added up to the conclusion that 'cropping pattern trials were wrong for subsistence upland farmers . . . After three years farmers had adopted only one maize variety, one peanut variety, and narrower gabi spacing. We judged the research an inefficient use of scarce resources (Lightfoot et al, 1985). We returned to an intuitive start and asked, what are the farmer problems and what resources are available to solve them? . . . the project had to develop new research techniques to get better answers'.

'Without doubt, the single most important reason for the failure of our early work was a lack of farmer participation in deciding the research agenda. Thus we started to devise mechanisms to find out what farmers were doing in terms of indigenous research and what problems they would like us to help them solve.'

In four of the six villages where the project was working, new directions for research developed around problems identified by farmers. One of these villages was Gandara where the invasion of fallow land by cogon grass was seen as critical. The extended surveys and discussion which led to this problem definition and the suggested solutions were described earlier in this book by Lightfoot et al (section 2.7). In other villages, problems identified by a similar process of survey, discussion and mapping included soil degradation and labour-use.

Another point of entry for the researcher was the experimentation in which farmers were already involved, for example, in assessing crop varieties, or growing upland rice on acid soils, or using local lime on some of these acid soils. It seems that there were two villages where this approach to on-farm research was important.

Yet another approach tried was a 'community based' project involving research into the planting of leguminous trees (mostly *leucaena*) as contour hedgerows in hillside farming areas. This work included farmer training with the new technology, followed by hedgerow planting by farmers with staff helping and monitoring progress. In one village, Villaba, neighbouring farmers came to learn about *leucaena* hedgerows from the initial farmers and after some informal discussion and demonstration, they returned to their farms and started using the method. So an informal farmer-to-farmer extension process began.

Summing up their experience of new approaches to doing systems experiments with farmers, Repulda et al (1987:9) conclude:

It was not until the blinkers of component and cropping trials had been removed that we saw the new and needed farmers' research questions (and) learned that farmer participation does not mean asking farmers to approve our experiments, but eliciting their experiments and designs. Often farmers were more prepared to go into the unknown than conservative researchers. They wanted to find out now; they do not want us to spend five years on the experiment station first. We learned that holistic does not mean redesigning the whole farm at once, but seeking interactions within the whole farm system. Lastly, we learned that farmers were much more interested in and got more involved in experimenting with solutions that lay within their own capacity than experimenting with conventional high-input solutions.

Such lessons seem simple and obvious once they have been learnt. But the diverse approaches of farmer participatory research conflict with conventional scientific training and practice. Institutional barriers impede their acceptance and adoption by scientists. It is not enough to recognize the power and validity of these approaches. It is also necessary to create institutional conditions in which they are feasible, supported and rewarded. And it is to these institutional and practical policy questions that we now turn.

Notes

1 Based on discussions in the ITK study group and informal comments by Anil Gupta, Roland Bunch, Lori Ann Thrupp and Ed Barrow.

PART 4

Institutions and practical change

Introduction

The final part of this book addresses the implications for institutions and for action of putting farmers and farm families first. Supporting farmers' innovations, learning from them, giving their agendas priority and spreading farmer participation in agricultural research, are all part of a 'farmer-first' mode of technology development. This complements, but contrasts with, that of transfer-of-technology (TOT). The institutional changes required include transformations of attitudes and behaviour and linking informal with formal R&D. In India, as KV Raman shows for training and interactions of scientists with farmers and as NK Sanghi shows for programmes and practices to bring scientists and farmers closer, efforts in these directions have made some progress in spite of institutional difficulties.

For the future, Robert Chambers outlines farmer-first roles, where outsiders as consultants support farmers' analysis and experiments, and search for and supply what farmers want and need. Changes in institutions, incentives and methods are implied, with action by a plurality of individuals and organizations in mutual support. Those who pioneered on these professional frontiers, and who improve and spread farmer-first approaches stand to gain the reward of seeing that poor farm families are truly served.

4.1 Context and change

IDS WORKSHOP[1]

The evidence and analysis in this book present elements of a 'farmer first' approach for agricultural research and development, complementary to that of 'transfer of technology' (TOT). Some of the basis of this approach is recognition of what Stephen Biggs (1978, 1979, 1981, 1987a) has called informal R&D (Research and Development), meaning the R&D conducted by farmers, artisans and other local people. Part 1 added to the evidence of the existence, relevance and vigour of informal R&D and of farmers' own innovations and experiments; and Roland Bunch in Part 2 argued the need to strengthen farmers' experimental capabilities and to enable them to innovate in a self-sustaining manner.

Formal R&D refers to the disciplinary activities and procedures of conventional research, as taught at agricultural universities and practiced in government and international agricultural research organizations. They include single commodity programmes, on-station seed-breeding and agronomic research. The formal and informal systems have been seen as

165

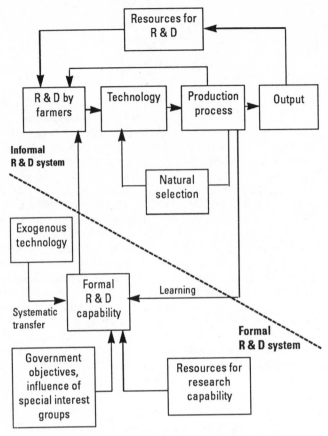

Figure 4.1: *A combined formal and informal agricultural R&D system*
Source: *Biggs and Clay, 1983*

analogues as illustrated in Figure 4.1. They differ in that the formal system is centralized, powerful and visible, while the informal is dispersed, weak, and hard for professionals to see.

Programmes which move the two closer together have started with the formal system and sought to extend it to the farm and sometimes to the farmer and farm family. At its simplest, on-farm research has been a move of scientists' experiments from research station to farmers' fields. But as we have seen, some researchers have gone further, to on-farm research under farmer management on problems which farmers identify as priorities. Such practices in on-farm research have given rise to differing degrees of participation and different relationships between researchers and farmers, which have been distinguished by Waters-Bayer (1989) and Biggs (1988).

Waters-Bayer separates out five main types of plot or field trials:

- scientists' on-station trials
- scientists' on-farm trials

166

- farmers' on-farm trials
- farmers' participatory trials
- farmers' informal trials

The first three are in the TOT mode, with the trials dominated by scientists, but with increasing involvement of farmers; in farmers' on-farm trials the innovation to be tested and the trial design are determined by scientists, but farmers make the management decisions. In contrast, farmers' participatory trials are in a farmer-first mode, with the questions to be investigated determined by farmers rather than scientists, who serve as advisors; and farmers' informal trials are informal R&D.

Biggs distinguishes four types of relationship between scientists and farmers in on-farm research:

contract: where the involvement of farmers is minimal. The scientists use farmers' resources, mainly land, for researcher-designed on-farm trials.

consultative: where scientists may interview and consult farmers about their problems at the start, and then decide on priorities and design of trials and surveys. This mode usually has a sequence of stages, such as diagnosis, design, technology development, testing and diffusion. Farmers may be interviewed at the end to assess new technologies generated by scientists. The scientist is dominant and the relationship is like that of doctor and patient.

collaborative: where there is sustained interaction between farmers and scientists. Instead of the stages of research in the consultative mode, diagnosis and evaluation are continuous. The farmer is a joint collaborator.

collegiate: where the formal research system actively strengthens informal research at the farmer and community level, and enhances farmers' capacity to make demands on the formal system.

Of these, contract and consultative are TOT, collaborative is borderline, indicating that we are dealing with a continuum, and collegiate is in a farmer-first mode. In Waters-Bayer's farmers' participatory trials and Biggs' collegiate relationships, it is farmers who dominate in identifying priorities and choosing treatments, and it is the farmers' evaluation which is the test of technology.

The distinction between TOT and farmer-first can be further illustrated by Biggs' analysis of the linkages between formal, institutional, R&D, informal R&D and resource-poor farming (Figure 4.2). In its classical form, TOT stresses direct linkages from formal system to farming system. Farming systems research is strong on the linkage in which understanding of the farming system is gained by the formal system, but tends to neglect informal R&D. The farmer-first approach, in contrast, makes informal R&D central, and stresses linkages between informal and formal systems.

TOT and farmer-first are paradigms, in the sense that each is a coherent and mutually supporting pattern of concepts, values, methods and action. They should be complementary, not alternative. On-station and in-laboratory agricultural research will always have a part to play. There will continue to be gains for much of the third – complex, diverse and risk-prone – agriculture, from commodity research, for example breeding for

167

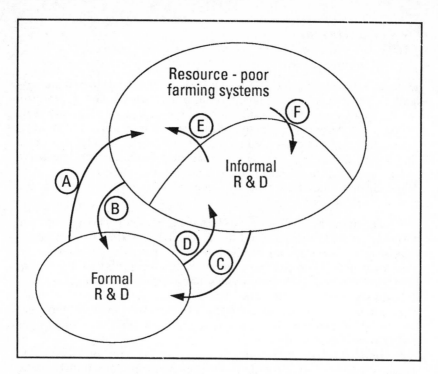

Figure 4.2: *Linkages of resource-poor farming systems with the informal and formal kinds of R&D. Linkage A, direct from formal R&D to farming system, is stressed by TOT approaches, although in practice the linkage is usually more like D, because farmers adapt as they adopt. Linkage B is stressed by Farming Systems Research, in which there is direct learning from resource-poor farming systems by practitioners of R&D. Both these approcahes tend to neglect informal R&D. However, farmer-first approaches make informal R&D central, and stress linkages C, D, E and F.* Source: *adapted from Biggs (1987).*

disease-resistance, drought-tolerance or early maturing, though these gains will be individually smaller and less widely applicable than the green revolution breakthroughs. There will always be aspects of farming and farm problems where scientists' skills will have a comparative advantage, as in diagnosing and treating the minute and microscopic in plant and animal pests and diseases, in finding trace element deficits, or in biotechnology. There will always be a role for specialized scientific services for farmers, not least in the prevention, diagnosis and treatment of pests and diseases.

That said, the biases supporting the normal TOT processes are so strong, and so mutually reinforcing, that they have to be recognized and resolutely reversed if farmer-first approaches are to achieve even a fraction of their potential. Recognition and reversal are not easy. Textbooks, curricula, teachers, trainers, professional peer groups, journal editors, and incentives and rewards in professions and bureaucracies – these are all weighted towards TOT. Moreover, it is personally gratifying to believe that one's knowledge is superior, and some consider it demeaning to learn from and with farmers, the more so if they are poor. TOT is defended by personal self-esteem and convenience, by professional incentives, and by the centralizing, standardizing and simplifying tendencies of bureaucracy.

TOT has also been defended by the efforts made to extend it to accommodate criticism and to move towards farms and farm families. Farming systems research has made a huge contribution to understanding. It has been a response to the complexity of farming systems, and has learnt from farmers; but often the learning has been transferred as information to 'our' knowledge, for analysis by 'us' and then transfer of recommendations back to 'them', still in the TOT mode. Similarly, most on-farm research, as contributions to this book have pointed out, has been in the researcher-initiated, researcher-managed mode, a transfer to farmers' fields of practices and methods from research stations.

Nor have even these extensions of TOT been easy to establish and maintain. One finding of the major ISNAR study in nine countries of on-farm client-oriented research has been the difficulty of sustaining client-oriented programmes after their initial phase (Merrill-Sands 1988). Like a magnetic field, the pull of the normal is always there and reasserts itself once countervailing forces weaken.

Much of the challenge presented by this book is thus institutional. It is not just for individual scientists to change, but for institutions in the formal sector to tolerate and support change and to spread it. This applies to all the major institutions in agricultural research, including:

- the international agricultural research centres (both members and non-members of the Consultative Group for International Agricultural Research)
- national agricultural research systems (NARSs)
- agricultural universities, and agricultural faculties in other universities

For its part, agricultural extension is usually TOT by definition. But it too could play a part in the farmer-first mode, linking with research, if effective reversals of learning and communication could be achieved and sustained.

To make headway against professional and bureaucratic inertia in large organizations such as these can be a formidable task. But that a start with progress can be made even in the very large research organizations found in India is shown by the two papers which follow.

4.2 Scientists' training and interactions with farmers in India

K V RAMAN

Problem identification as a gap in scientific training

The training that research scientists receive at university helps them to develop into excellent subject matter specialists. However, one component

169

of training usually missing is problem identification. In a survey of a large number of research projects where an attempt was made to analyse the original source of ideas, it turned out that more than half were undertaken by the scientists essentially as an extension of their graduate work or had arisen without defined objectives. Only 15–20 per cent of the projects had been developed after a field study that attempted to identify specific problems that required solution.

This study was extended more recently by the National Academy of Agricultural Research Management at Hyderabad by an analysis of over 300 research projects at research stations within the agricultural university system in India. It was gratifying to find that there was a shift in the source of ideas for generation of projects, with more projects now arising from interaction with farmers and extension workers.

The lack of problem orientation in the earlier study may be related to the urban background from which many researchers now come. A few decades ago, most of those who took to the study of farming science came from a rural background and had some prior experience of agriculture. Thus they had some awareness of farming problems, even if they were not taught formally to identify them, but now with many students coming from urban areas, there is a missing link of living and working in a rural environment.

In response to this, the Andhra Pradesh Agricultural University at Hyderabad introduced a novel scheme of providing a learning experience to students from farmers. This programme is known as Rural Agricultural Work Experience (RAWE) and its objectives are, first, to provide an opportunity for students to live in rural areas; second, to help them gain first-hand experience of agricultural technology in farmers' fields; third, to ensure that students appreciate the constraints limiting the application of new technology; fourth, to develop communication skills; fifth, to help students to develop a constructive attitude towards the farming community.

The duration of the programme is one semester, that is, about five months. Batches of between two and four students each are seconded to selected villages in the vicinity of a research station and are attached to one or two host farmers. While living in the village and interacting with the entire community, with the host farmer in charge, the students not only participate in the day-to-day agricultural operations, but also study the local agricultural situation, crops and cropping patterns and problems related to such matters as water management or disease and pest control. During this period, the students maintain a close rapport with the research station and its scientific staff and get there such assistance as they may need in identifying problems and finding solutions for them. The focus of this training programme is 'learning from the farmer' rather than providing the farmer with technical advice and expertise.

The programme has been in operation at the Agricultural University for the last six years. At first there was considerable resistance from the students, as they were required to live in the rural areas in conditions to which they were unaccustomed. However, this resistance rapidly wore off as the value of the learning experience came to be appreciated. Students even began to introduce new elements into the programme such as organizing field days, putting up small exhibitions and interacting with farmers during Panchayat meetings. Realizing that village discussions on farming problems mostly took place in the evenings, they understood that it was necessary to live in the village to participate in these discussions. Periodic visits to farmers during the cropping season when they are busy in the fields do not provide opportunities to listen to the farmers' point of view or learn from their technical know-how.

At first female students, were excluded from the programme as it was felt that they might face many social problems living in the villages. However, they were later included at their own request. Now instructors feel that they do even better than the male students since they are able to establish instant rapport with farming women who have a dominant role in many agricultural operations.

The host farmers have also taken to this programme well. During critical operations, such as spraying for disease or harvesting, an extra hand at the farm level is always welcome. The supervisory staff feel that students benefit a lot by this kind of exposure as it makes them appreciate the constraints which prevent many 'scientific recommendations' from being implemented at farm level.

Based on this very positive feedback and, realizing the immense potential such programmes will have on improving standards of agricultural education, several other universities have now considered introducing similar programmes as part of their curriculum.

Pre-service training for ARS scientists

The Indian Council of Agricultural Research (ICAR), as the chief federal agency co-ordinating agricultural education, research and first-line extension activities, has over 6,000 scientists[2] in about 50 research institutes and national centres working on different crops, commodities and animal species. These scientists are recruited into the service in 57 different disciplines and have minimum educational qualifications of Master's degrees. Realizing the need for a training in *research project management*[3] for these scientists, the Council set up an Academy in 1976 to give them a basic orientation training. It is mandatory for all the scientists joining the Indian Agricultural Research Service (ARS) to take this 'foundation training' course of five months duration within a period of two years after joining the service.

This course is conducted in three phases. In the first, the trainees spend six weeks acquiring a perspective of the Indian agricultural scene and learning about national priorities in agriculture and the organization of the National Agricultural Research System. They also tackle basic themes related to project development and management, including identification, formulation, implementation, monitoring and evaluation of research projects. In the next phase, of two months duration, known as the Field Experience Training Programme, the trainees are attached to an agricultural research station located in a rural area. During their stay, they do some survey work and also formulate a research project in their field of specialization after interacting with farmers and extension workers, attempting to identify 'problems of relevance'. In the third phase of the course, they return to the Academy for a further programme of six weeks, when they present their research projects, exchange experiences and undergo training in the areas that will equip them to become better research workers. These include studies in the communication process and treatment of research findings.

We would particularly like to share our experience of the Field Experience Training phase, which we consider to be an important learning experience for the scientists. Among the objectives of this part of the course, there is an aspiration to make the scientists aware of the technology development process and also to study the mechanism of linkage between agricultural research and extension and its role in technology transfer. The programme is intended to help participants to understand the existing rural development machinery at village, block and district level and by exposing the scientists to the rural setting, to make them understand the requirements of the clientele with respect to agricultural research and the socioeconomic constraints affecting the adoption and diffusion of technology.

What these objectives mean in practice is that the scientists are sent to the regional research stations belonging to the agricultural universities. There, they have discussions with staff concerning the current development process and they undertake some survey work, both among research station scientists and also at village level, where they interview 50 or more farmers from at least two villages in order to identify major production constraints affecting crops or animals in the region. They then develop proposals for research projects with a problem-solving emphasis and an interdisciplinary approach. Care is taken to ensure that the trainee scientists posted to a particular research station belong to different disciplines so that when they travel and work together, their perception and approaches to the problems will vary and they will learn from each other. This sensitizes the trainees to the need for developing multidisciplinary research projects.

The survey data collected by the trainees from scientists in the research stations includes information on the sources of ideas for technology and development as well as criteria used by them for setting priorities in selection of projects. It is apparent from the results (Table 4.1), that contact with farmers and meetings attended by farmers and extension workers are now an important stimulus for project development.

For a large number of participants who come from urban backgrounds and have never experienced the problems of rural life, the field experience training has proved exhilarating and useful. Talking to the farmers exposes them at first-hand to the utility or otherwise of some of the existing technologies and makes them aware that ivory-tower research may not solve many of the problems of the rural community. It also exposes them to the resource constraints of the farmer and makes them realize that solutions to their problems should not be 'resource-intensive'. Many of them also come to realize that the first stage in the project cycle, problem identification, is the most difficult one, requiring a thorough understanding of the crops and cropping system as well as a shrewd insight into the entire agricultural development process of that region. Many of them also realize

172

Table 4.1   Sources of ideas for technology and development given
            by scientists in research stations

Sources of ideas for technology development	Percentage of scientists responding
1 Workshops/meetings/conferences (often with farmers or extension workers participating)	56
2 Previous research projects	55
3 Journals/books	47
4 Contact with the farmer	41
5 Field survey	41
6 Contact with the extension worker	30
7 Personal background	22
8 Others	20.5

(Questions asked for multiple responses where this was appropriate, so
the percentages total more than 100.)

how their colleague scientists in the research stations need a more
analytical approach to the various components of the project cycle.

The National Agricultural Research Project

The National Agricultural Research Project (NARP) is an innovative
venture within the National Agricultural Research System in India designed to assist the agricultural universities to conduct need-based, location-
specific and production-oriented research. The entire country is divided
into 126 contiguous agro-ecological zones, identified on the basis of
similarities in soils, climate and ecological conditions. A major regional
research station in each agro-ecological zone, with a group of scientists
from different subject disciplines, is developing a focus on problems of
local and regional interest.

If agricultural research is to benefit small farmers, their resources,
environment, knowledge and attitudes must be taken fully into consideration before any project is even conceived for that region. Thus one
feature of NARP is a basic survey of resources and constraints in each
region by a multidisciplinary team. The aim is to ascertain the potential
flexibility of the farming system by interacting with the farmers and
extension agencies through formal and informal surveys.

The information obtained, pertaining to different farming situations with
specific production constraints, takes account of such factors as climate,
soils, crop varieties, cropping systems, water management, plant protection and socio-economic conditions. It is all documented in a 'status
report' for the region which is intended to serve as a resource inventory

and hence as the basic document for planning research in the region. However, studies undertaken by the National Academy of Agricultural Research Management show that some of the earlier status reports were not as useful as they should have been since they did not give appropriate descriptions of the various farming systems prevalent and did not have adequate regard for farmers' indigenous technical knowledge. Later reports, prepared on the basis of a changed format, are considerably improved.

Another limitation of the earlier work was that farmers and extension workers were not much involved in the identification of problems, often because it was thought that their approach would not be scientific. Realizing the futility of this argument, and appreciating the need for farmers to be involved in the planning cycle, NARP has evolved a system for their participation, beginning with diagnosis of problems and continuing with farmer-managed trials.

One feature of these arrangements for participation is that zonal workshops are held twice a year, before the kharif and rabi seasons. Apart from generating research ideas through discussion between scientists, development agencies and farmers, these workshops review research and field testing results in relation to the needs of the zone. The number of workshops conducted by the different centres in 1985 and 1986 varied somewhat since this philosophy of interaction with farmers and development agencies is still new and has taken time to sink in. However, in older projects, workshops were conducted more regularly and it was noted that the participation of the farmers and extension workers was also becoming more meaningful and useful.

Table 4.2: Details of projects conducted by five research stations during 1985 and 1986

Research stations	Percentage of projects with multi-disciplinary approach	Number of villages adopted	Number of on-farm trials or demonstrations per season
1 Bawal, Haryana	26	5	25
2 Tirupati, Andhra Pradesh	32	2	4
3 Pillicode, Kerala	56.5	4	–
4 Bijapur, Karnataka	40	2	30
5 Sriganganagar, Rajasthan	12.5	–	20

Source: Evaluation studies conducted by the National Academy for Agricultural Research Management

174

Another change is that while at one time most projects were discipline-based, NARP has specifically emphasized the multi-disciplinary approach in problem identification. Even so, multi-disciplinary projects are still in the minority at most research stations as Table 4.2 illustrates. One of the problems may be the organization of the divisions themselves in the research stations which were discipline-based and, therefore, wanted to have an identity of their own.

On-farm trials provide a link between farmers and scientists, ensuring the relevance of on-going research to the farmers' communities. They can help remove barriers to the adoption of station-developed technologies. Under NARP, testing and evaluation of the technologies developed at the research station are undertaken at the farmers' level to provide necessary feedback and to ensure that research programmes become increasingly responsive to farming problems.

At most research stations, there was no systematic on-farm work prior to the introduction of NARP. However, regional research stations, developing technologies suitable for their locality, now adopt villages for participation in trials, and have conducted many on-farm experiments (Table 4.2). It is apparent, though, that the concept is just catching on and it may have a long way to go before it is effectively implemented in many places.

All the above measures are designed to provide the scientist with opportunities to interact with farmers, understand their constraints, and formulate action-research programmes of relevance to the region.

After only five years since the project began, there is evidence of a distinct improvement in the extent of on-farm research. There is also an increase in the practice of inviting farmers to visit research stations for training and discussion with scientists concerning the latest technologies.

At the same time, as we saw earlier in the paper, the training of researchers is changing in ways which bring them into closer contact with farmers. There is thus a promise and a hope of institutional development at several levels which will draw research increasingly out of the ivory tower and onto the farmers' fields.

4.3 Changes in the organization of research on dryland agriculture

N K SANGHI

Research-extension linkages

During the last two decades, much has been achieved in research on dryland agriculture in India, but it has been observed that 70 per cent of the new technologies developed are confined to research stations and their annual reports. Whilst one response has been the suggestion that research stations need to be more relevant to the felt needs of farmers, the initial,

175

more widespread response was to intensify the extension programme and provide more training and more timely inputs for the farmers.

In recent years, the intensification of extension work has been made chiefly through the well known training and visit system. Analysis of the existing situation in dryland areas of Andhra Pradesh has, however, shown that the new technologies for drylands have still not been able to make much contribution to productivity levels of the majority of rainfed crops, even in areas where intensive extension has been carried out for almost a decade.

Simultaneously, a great deal of effort has also been made towards re-orientation of research. In the case of dryland agriculture, this has been done mainly through the introduction of 'interphase projects' (ie, projects in between the research and extension programme). Such projects have been designed to test pre-release recommendations under the real farm situation, so that verified technology could be promoted for large-scale adoption. They were also expected to provide feedback to the scientists for further refinement of technology to suit the socio-economic conditions of the farmers.

The 'interphase projects' have had the great merit of influencing both extension work and the research programme (Friesen et al, 1982; Sanghi et al, 1983) but, in a number of cases, feedback has been weak due to lack of proper interaction with the farmers. Hence there is a need to improve the methodology further so that farmers in these projects can participate not merely as beneficiaries but mainly as co-research workers, so that they may have some control over the situation.

The old arrangements, typical of the 1950s and 1960s, whereby research results went directly to the extension service for large-scale application in farmers' fields, seem to have worked satisfactorily for irrigated agriculture or for resource-rich farmers since the farming situation at the research station and in the farmers' fields did not differ too much. Such a set-up, however, was inadequate for dryland agriculture because of large and obvious differences between research conditions and farmers' fields.

A striking improvement was, therefore, made on a trial basis in the All-India Coordinated Research Project for Dryland Agriculture. From inception of the project, during the early 1970s, the concept of a pilot development project (PDP) was introduced as an interphase between the research and extension programmes. Under this set-up, an area of about 4,000 ha was taken in selected villages adjacent to each of the 23 research stations for dryland agriculture in India. Each year the project was operated in about 800 ha so that the complete area could be covered in five years. These projects were operated by the full-time extension staff under direct guidance of the scientists from the associated research stations and were able to demonstrate the potential of new technology in a very effective manner.

A critical review of the project during the mid-1970s has, however, revealed that although the farmers adopted the improved technology during the project implementation years, they reverted to their traditional systems as soon as the project moved on, even if only to an adjacent area (Friesen et al, 1982).

At that stage there were was a general tendency to blame either the technology or the extension efforts, depending on whether scientists, extension staff or farmers were interviewed. It was, however, realized that the methodology used in these projects suffered from certain inherent limitations (Krishnamoorthy, 1975). By insisting that a 'full package' of measures was implemented, scientists made no allowance for understanding whether there was any weakness in certain components of the technology. Similarly the subsidy on inputs did not clarify whether the farmers' adoption during the project implementation years was due to strength of technology, or due to the attraction of the subsidized inputs.

With this kind of experience, it was considered appropriate to modify the interphase project in order to get a clear understanding of the above aspects. Hence the concept of an operational research project was introduced in dryland agriculture during the mid 1970s. This type of project was operated only by scientists appointed exclusively for this purpose and it was run as an integral part of the work of research centres for dryland agriculture.

Lessons learned from operational research projects

Operational research projects have been very successful in revealing the wisdom of many traditional practices and in providing feedback to research scientists regarding weaknesses in certain components of their technologies.

It has been increasingly realized that dryland farmers in India have, through long experience, evolved many useful practices. Examples include methods of avoiding pests and diseases (through choice of proper sowing time and duration of crop variety); in situ conservation of moisture (through suitable land treatments); minimization of risk (through intercropping, agroforestry and contingency cropping); and reduction in the cost of weeding (through early interculture operations). Details of many of these site-specific traditional practices have been reported separately (Friesen et al, 1982, Sanghi et al, 1983). It is clear, then, that we should be ready to learn from farmers. It is also clear that there is no need to make a total change in the system while introducing improved technology. Instead, a mixed approach of bringing in some new practices while retaining those that are sound in the traditional system may be economical and may also lead to better involvement of the farmer.

The dryland operational research project has also made a useful contribution in the critical evaluation of each component of the new technology. Nearly 25 per cent of the research station recommendations have been found to be unsuitable and have been referred back for further refinement. Out of the remaining recommendations, about 30 per cent have shown high profitability and a rapid rate of diffusion among the farmers, whereas the rest of the recommendations did not spread in spite of their successful performance in the verification trials.

Despite these interesting results, operational research projects proved to have a number of limitations. One was that they could not identify site-specific, non-technical constraints. A second was the temptation confronting scientists while testing the technology to achieve success by taking control over the trials even though these were conducted in the farmers' holdings.

A third limitation was that some production problems were not tackled by the scientists at the research station. This was either because the production problems were too difficult (ie, requiring long-term research) or because they appeared trivial or of limited application.

Thus the existing methodology in the operational research project could not provide the farmers with significant control over research priorities, even though it did give clear feedback. At the same time, there was no in-built mechanism to involve farmers in making modifications in the technology at the village level.

Farmers as co-researchers

Some of the scientists working in operational research projects have attempted to overcome these limitations in the research format by investigating whether and how farmers can participate as co-research workers, even to the extent of modifying technology. One example comes from a project located on the red soils of Hyderabad and concerns placement of fertilizer, which is considered to be critical under rain-fed conditions. Keeping in mind the available draught power, the scientists have evolved a series of bullock-driven, seed-cum-fertilizer drills for varying soil types and rainfall situations. In Telangana region of Andhra Pradesh, the *Fespo* plough is an improved device of this kind for red soils where the country plough is used and *Enati gorru* is an improved device for black soils where the country *gorru* (three-row seed drill) is used.

These implements have shown the desired performance, but farmers have preferred to simplify them in order to reduce the cost and also to make them more convenient to use. It was interestingly observed that the red soil farmers could design four alternative methods to achieve the fertilizer placement. All of the alternatives could obtain nearly the same results as that of the original equipment whereas the cost was significantly reduced and locally available material could be used (Table 4.3).

Another example concerns the broad bed and furrow method of water conservation developed by ICRISAT in the mid-1970s. While implementing this practice under village conditions, researchers realized that it could not be adopted by the local farmers due to the high cost of equipment and its inconvenience in operation.

Analysis of traditional practices in one area later revealed that in the case of castor crop farmers follow the same principle of moisture conservation but in a modified manner (Sanghi et al, 1983). They leave the plough furrow open after the sowing operation and also make new furrows during interculture operations to hold back water after heavy rain. Subsequently, systematic trials at research farms have revealed that the farmers' method of moisture conservation with castor is almost as good as the bed and furrow system and requires no extra equipment or any special efforts.

178

Table 4.3: Alternative methods for fertilizer placement in red soils of Hyderabad developed by farmers, compared with the Fespo plough developed by researchers.

Item	Cost (Rs)	Description and method of operation
1 Fespo plough	200	The speed of operation is about 50 per cent faster than traditional method, provided land is prepared in advance.
2 Plough attachment (metal set)	80	It can be fitted to the existing *desi* plough
3 Two tubes attached behind country plough	10	Seeding and fertilizer by *pora* so that intercultural operations are performed efficiently.
4 One tube attached behind country plough	5	In this case, seeding is done by usual *kera* method.
5 Country plough	0	Seed and fertilizer falls in the same furrow. Only one level of nitrogen can be used at sowing (eg, 10–15 kg N/ha).

Notes:
Desi plough: country plough
Pora method: the seed and fertilizer are drilled through a tube
Kera method: the seed is dropped by hand in the furrow opened by country plough (without using any tube)

Other examples of farmers modifying technology developed by researchers, or using alternative technologies of their own, include a modified bunding system for soil conservation and methods for controlling pests and diseases in sorghum (Sanghi et al, 1983). However, there are institutional difficulties in formally recognizing practices which emerge due to modifications at village level. Normally, all technologies are evolved and formalized by scientists working at research stations. Field trials in the farmers' fields are only supposed to be a means of verifying the findings and choosing items to remake packages. However, it is now being realized that a distinction needs to be made between research results and relevant technologies. In other words, sufficient evidence is now available to show that relevant technologies can be generated through a joint intervention by scientists with extension workers and farmers. If this assumption is correct, a working arrangement has to be evolved to formalize those practices which are modified at village level so that they may be included in regular development projects.

Issues for consideration

The major question arising from these experiences is institutional, but at a different level from those considered in section 4.2. There we saw how major research institutions in India are pressing for more on-farm research and how, indeed, there is an encouraging trend in that direction. Here we have to ask about the situation where on-farm research is accepted. How can farmers then be given more control over research priorities? How can they be offered opportunities for modifying technologies to suit their local situation? Only tentative suggestions can be made about this. However, two points with less far-reaching implications can be tackled, concerning the temptation for scientists to push for success in on-farm trials, and the attraction toward subsidies among farmers.

The temptation-for-success factor would be less if there were a clearer distinction in on-farm research between farmer-managed and researcher-managed trials. Scientists would not intervene in the former even when it seemed to them that a small modification could correct a trial that was going badly. A further check on researcher-bias could be provided by conducting constraints research ie, after trials are complete, asking farmers for views on the technology, and their reasons for not adopting it and where necessary releasing the constraint on a trial basis to see whether that then leads to adoption of the technology.

An example concerns field trials on castor held in the Hyderabad operational research programme in 1979. Improved equipment for fertilizer placement was introduced and demonstrated successfully but subsequently the farmers did not adopt it. The follow-up survey revealed that high cost and lack of availability of equipment were the main reasons for non-adoption. Further constraints research, however, showed that this analysis was biased, because hardly any of the 10 farmers who were provided with the improved equipment free of cost adopted the improved method of seeding. Such experiments have clearly illustrated the value of an additional step, namely constraints research, in avoiding subjective analysis.

Attraction toward subsidy is the other factor which leads to a subjective assessment of technology, particularly during the project implementation years. The problem has been minimized to a large extent in a watershed development programme near Gulbargah in Karnataka by introducing the concept of 'risk cover'. Improved technologies for seven different crops were introduced during the Kharif season in 1986. Before the implementation of the programme, the details of various components of improved packages for each crop were thoroughly discussed with the participating farmers. After the discussion, the content of the package for each crop was classified according to two categories: *non-controversial* practices (D1) and *controversial* practices (D2).

Non-controversial practices were those components of the package about which the farmers were convinced. In the subsequent trials, these components were provided to the farmers at their full cost but farmers were offered risk cover for their expenditure in case production on the trial plot was low as compared with a control plot. Claims had to be lodged with

180

the implementing agency before harvest so that the agency had the option of itself harvesting the plots and checking the farmers' claim.

Controversial practices were those components of the package which researchers felt would be profitable whereas farmers thought otherwise. In such cases the cash inputs for trials were paid by the implementing agency. Experience with this approach has revealed that the work load for conducting systematic trials can be reduced considerably when the non-controversial components of the package are separated at the beginning. The most significant advantage of the 'risk cover' approach was that it created a built-in mechanism for getting feedback regarding the assessment of the technology.

The farmers did not hesitate to point out even a small weakness in the package since initially they had invested their own money for purchase of the inputs. Since the claim for refund had to be made before the harvest of the crop, the concerned scientist (who had evolved the recommendation) could either justify the recommendation or accept the farmers' complaint.

Field experience during 1986 in Gulbargah watershed has shown that out of packages for seven different crops, farmers approved the sorghum and groundnut packages (which covered about two-thirds of the total programme). The new practices for greengram and blackgram could not be properly assessed during 1986 because of severe drought. Regarding sesamum, the improved variety showed lower yield potential than the local one. With pearl millet, the recommended hybrid showed no grain filling in the primary ear-head although other heads were properly filled and with paddy, fertilizer use was not economical with the local variety.

The problem of giving farmers greater control of research remains. However, some of the procedures just described may help in this respect, including risk cover for non-controversial (D1) technologies and farmer-managed trials for more uncertain (D2) technologies. However, it is still necessary to devise ways whereby farmers might participate more fully in the planning stage of interphase projects. The formalization of technologies which are modified at village level through the direct or indirect involvement of farmers is another institutional issue which remains unresolved.

4.4   Reversals, institutions and change

ROBERT CHAMBERS

Farmer first and TOT

The new behaviours and attitudes presented by the contributors to this book conflict with much normal professionalism and with much normal bureaucracy. Normal professional training and values are deeply em-bedded in the transfer-of-technology (TOT) mode, with scientists deciding research priorities, generating technology and passing it to extension agents to transfer to farmers. Normal bureaucracy is hierarchical and centralizes, standardizes and simplifies. When the two combine, as they do in large organizations, whether agricultural universities, international agricultural research centres, or national agricultural research systems (NARSs), they have an impressive capacity to reproduce themselves and to resist change.

But to serve well the resource-poor farm families of the third – complex, diverse and risk-prone – agriculture with which much of this book has been concerned, requires these 'normal' tendencies to be reversed: for farmers' analysis to be the basis of most research priorities, for farmers to experiment and evaluate, for scientists to learn from and with them; and for research and services to farmers to be decentralized, differentiated and versatile.

The difficulty of effecting major changes and reversals in large organizations underlines the importance of seeing what changes of behaviour and attitude are required, what institutional conditions are necessary for them to be sustained and spread, and how these might be achieved. To do this, we need to outline in more detail the contrast between TOT and the farmer-first approach and methods represented in this book (see Table 4.4).

Table 4.4: Transfer-of-technology and farmer-first compared

	TOT	FF
Main objective	Transfer technology	Empower farmers
Analysis of needs and priorities by	Outsiders	Farmers assisted by outsiders
Primary R&D location	Experiment station, laboratory, greenhouse	Farmers' fields and conditions
Transferred by outsiders to farmers	Precepts Messages Package of practices	Principles Methods Basket of choices
The 'menu'	Fixed	A la carte

With farmer first, the main objective is not to transfer known technology, but to empower farmers to learn, adapt and do better; analysis is not by outsiders – scientists, extensionists, or NGO workers – on their own but by farmers and by farmers assisted by outsiders; the primary location for R&D is not the experiment station, laboratory or greenhouse, necessary though they are for some purposes, but farmers' fields and conditions; what is transferred by outsiders to farmers is not precepts but principles, not messages but methods, not a package of practices to be adopted but a basket of choices from which to select. The menu, in short, is not fixed or table d'hôte, but à la carte and the menu itself is a response to farmers' needs articulated by them. All this demands changes in activities and roles.

Farmer-first activities and roles

Contributions to this book show farmers carrying out or participating in various activities which in the TOT mode are conducted only by scientists. Three of these, again and again, are analysis, choice and experiment. To support farmers in these activities generates and requires new roles for outsiders:

Farmers' activities	New roles for outsiders
analysis	convenor, catalyst, adviser
choice	searcher, supplier, travel agent
experiment	supporter, consultant

What these activities and roles entail can be illustrated by contributions to this book, supported by other sources.

(i) Analysis. Analysis by farmers takes many forms and can be promoted in many ways, involving outsiders to different degrees. In the examples in this book an outsider has often played a role, whether as questioner, convenor of a group, stimulator of discussion, or catalyst whose presence speeds up the process.

Analysis can be part of or generated by the use of a method. Some examples are:

- open interviews and iterative group conversations (Floquet, 1989);
- ethnohistory and ethnobiography (the biography of a crop, or of a person's experience of a crop, an historical analysis of the experience of a community, etc) (Box, Rocheleau et al);
- inspection and discussion: visiting trial sites, observing innovations, field days, and visits by farmers to research stations when they observe and discuss (Ashby et al; Norman et al);
- visual aids to analysis: seasonal and other diagramming (Conway), aerial photographs (Carson, 1987), systems diagramming on a board (Lightfoot et al), other uses of diagrams with and by farmers and communities (Kabutha and Ford, 1988; McCracken, 1988) and drawing maps (Gupta);
- eliciting clients' criteria and preferences, where individuals or groups (women, men, farmers etc) articulate their reasons for preferences, and then rank items according to them (Ashby et al; Chambers 1988);

183

- key questions and approaches to questioning: 'ways in' or 'points of entry' such as 'What would a desirable variety look like to you?' (Ashby et al, 1987:27), 'What would you like your landscape to look like in the future?' (Rocheleau), 'When you were a boy, what was the oldest variety of (a particular crop) that you knew about?' (Box) and 'Comparing agriculture practiced at the time of your father and grandfather with the agriculture practiced by you today, what are the major changes that have occurred?' (Gubbels, 1988);
- contrast analysis, where groups or individuals are asked to explain the contrasting conditions or behaviour of others, thus setting a frame of reference before analysing their own (Gupta)
- sequences of meetings and visits (Rocheleau et al, Mathema and Galt, Norman et al, Lightfoot et al, Repulda et al);
- innovator workshops where farmer innovators meet to discuss their new practices (Abedin and Haque, Ashby et al).

The role of the outsider is to elicit, encourage, facilitate and promote analysis by farmers, providing where necessary the stimulus, the occasion and the incentive for meetings and discussions. The outsider can take part, but does not dominate. Farmers' own analysis, criteria and priorities come first. Requests are generated for outsiders to search for what farmers want and need, and to provide them with choices or ideas for experiments to solve a problem or exploit an opportunity (Lightfoot et al., Repulda et al).

(ii) Choice. Choice by farmers is prominent in the farmer-first paradigm. It has two aspects. First, farmers' analysis generates an agenda of requests for information and material. Second, farmers need a range of choice, so that they can pick and choose to suit their conditions, extend their repertoire and enhance their adaptability. Norman et al note 'the technology assessment process in which a wide range of options are presented to a large number of volunteer farmers' (p. 141). To find and present variety and choices to farmers is largely a task for outsiders. Some examples are:

- providing farmers with varied genetic materials to test and appraise (Maurya, Rocheleau et al., Ashby et al, Norman et al);
- planting a variety of lines or species, to be followed by 'wait-and-see and pick-and-choose';
- issuing mini-kits of seeds and fertilizers to farmers for them to try out in various combinations;
- requiring nurseries, as with forestry in Kenya, to plant and provide a range of species, including a preponderance of indigenous species;
- transferring genetic material between regions, countries and continents, especially of non-cereal plants (multi-purpose trees, shrubs, grasses, vining plants, root crops etc) and livestock;
- transferring indigenous technical knowledge and practices between farmers in different regions;
- enabling farmers to travel, visit, see and learn for themselves the farming practices of others.

The role of the outsider, whether scientist, extensionist, or NGO worker, is to search for and supply the species, varieties, treatments, cultural practices, scientific principles, or combinations of these which fit and meet farmers' requests and needs. It may also be that of travel agent or tour operator, to arrange for farmers to visit research stations, other farmers, or other regions, to learn from other farmers and scientists and to widen their experience and options.

(iii) Experimenting. Experimenting by farmers has long been under-perceived. The professional world has been slow to recognize farmers' experimental inclinations and abilities (but see Johnson, 1972; Richards, 1985; Rhoades, 1987; Rhoades). Rhoades and Bebbington (1988) have identified three reasons why farmers experiment: to satisfy curiosity; to solve problems; and to adapt technology. As we have seen, their farming is both performance (Richards) and in a sense a continuous experiment: Hossain et al point out that farmers in Bangladesh are continually changing their cropping patterns (p. 35) and Juma puts it that 'a farmer is a person who experiments constantly because he is constantly moving into the unknown' (p. 34).

In the farmer-first approach, it is not packages of technology that are provided to farmers, but genetic material, principles, practices and methods for them to test and use. Genetic material can take many forms and may come from nearby, from other regions, or from other countries or continents. Similarly, principles can originate from different sources: in West Africa, the principle of alley cropping was taken from the research station and was adapted and experimented with by farmers (Sumberg and Okali); the principle of diffused light to inhibit potato sprouting in store originated with farmers in Kenya and was spread internationally and laterally to other farmers in many countries, who made their own applications with local materials to fit local farm architecture (Rhoades). Experimental principles and methods suitable for their conditions and needs can also be provided to farmers to improve their investigations and innovations (Bunch 1985, Bunch).

Farmers' experiments are, then, encouraged and supported by outsiders. This is close to Biggs' collegiate mode of farmer-scientist *interaction*. Farmers take part in design (Fernandez and Salvatierra), determine management conditions and implement and evaluate the experiments. They 'own' the experiments and the outsiders provide support and advice.

Evaluation of experiments is also by farmers and continuous. An authoritative World Bank publication (Casley and Kumar, 1987:116) has pointed out that it is often assumed that illiterate, tradition-bound farmers cannot assess the dynamics of change, but that their knowledge and judgments are in many instances more accurate than those of project staff. One of DM Maurya's criteria for assessing a line given to a farmer to try is whether other farmers ask for seed (per. comm.). It is farmers' judgments, interest and adoption that count.

Stimulating, servicing and supporting these farmers' activities – analysis, choice and experiment – requires reversals of normal and expected roles on the part of outsiders, be they scientists, extensionists, or workers in NGOs. This does not mean that they have to be purely passive catalysts. It would be as absurd for their ideas and knowledge not to be brought into play, as it

185

has been for those of farmers to be neglected. In raising questions, in providing tools for analysis, in presenting what they already know to be feasible and available choices, and in supporting and advising on farmers' experiments, they have a part to play. But their role is not that of teacher, of the bearer of superior modern technology, of the person who knows what is good for others better than they know for themselves. It is neither the role of traditional agricultural extension, nor that of normal agricultural science. An open, learning process approach is indicated, of a sort encouraged neither by the content of university curricula nor by the hierarchy and style of government bureaucracies.

For these changes and reversals of role to occur on any scale is not easy. It requires resolute changes in institutions, in incentives and in methods and interactions.

Institutional change

Unfortunately, normal bureaucracy tends to centralize, standardize and simplify, and agricultural research and extension are no exceptions. They fit badly, therefore, with the conditions of resource-poor farm families, with their geographical scatter, heterogeneity and complexity within any farm and farm household. In resource-rich areas of industrial and green revolution agriculture, production has been raised through packages, with the environment managed and controlled to fit the genotype. The third agriculture, being complex, diverse and risk-prone, requires the reverse, with searches for genotypes to fit environments. In industrial and green revolution agriculture, higher production has come from intensification of inputs and simplification and standardization of practices; in the third agriculture, it comes more from diversifying enterprises and multiplying linkages. Green revolution agriculture has been convergent, evolving towards common practices; the third agriculture often needs to be divergent, evolving towards a greater variety of differing enterprises and practices.

At first sight, then, the farmer-first approach appears incompatible with normal bureaucracy. But as contributors to this book have shown, reversals in government research organizations, though difficult to start and to sustain, are not impossible. Some contributors were working in special projects linked with NARSs; others were working in more normal conditions, as with the innovator workshops in Bangladesh (Abedin and Haque) and the distribution to farmers of advanced lines of rice in India (Maurya).

For the future, to achieve farmer-first reversals in national bureaucracies, especially NARSs, three aspects of management merit special attention: decentralization and resources; search and supply; and incentives.

(i) Decentralization and resources. Central controls need loosening if local actions are to fit diverse conditions. Centralized permissions for expenditures constrain flexibility. Centrally co-ordinated trials limit discretion and the ability to serve local priorities. When resources such as transport and money for travel are scarce, local discretion and control become more

186

important than ever. The essence of farmer-first approaches is to serve and support local diversity, with a reversal of demands on staff, the demands to come from farmers below more than from seniors above. Decentralization is difficult in normal bureaucracies. Central accountants fear loss of control over expenditures. Central officials fear loss of power and prestige. Reports are harder to collate and present, and work harder to supervise, when activities are varied. Methods are needed, and perhaps easier now with microcomputers, for valuing local diversity in staff activities in place of counting reported achievements of standard targets. For NARSs, the practical implications are to devolve resources and discretion more to the local level.

Freedom and means for staff to visit and spend time with farmers are crucial. For travel, something can usually be done quite simply. In the joint trek in Nepal, scientists walk together for days (Mathema and Galt). Foot, bicycle, horse and public transport can, variously, be used. For cost-effectiveness, though, other means of travel can be important, especially when distances are great and environments diverse. Unfortunately, access to transport and permission to use it are frequent problems, though less so with foreign-funded programmes. Travel and allowances can be high-profile privileges for which staff compete, jealously guarded and sparingly allocated by directors of institutes and heads of units. Worse, when revenue shortfalls or national policy reforms force cuts in recurrent budgets, staff are usually protected and it is other votes that suffer. Fuel, vehicles and nights out allowances are favourite victims. In Zambia, the Ministry of Agriculture's vote for petrol and maintenance had been reduced by 1980 to only one fifth of its 1973 level despite an increase in vehicles and staff (ILO, 1981:xxvi). Scientists can usually work with farmers close to their research stations and residences; but without hassle-free and adequate access to means for travel, it is difficult for them to work regularly and well with others further afield.

(ii) Search and supply. Search is neglected and rarely rewarded as a professional activity. This includes search for farmer-innovators and ex-perimenters, for genetic material, and for principles, practices and tech-nologies, whether locally, regionally, nationally or internationally.

Search is basic for meeting farmers' needs and widening their choices. In complex, diverse and risk-prone agriculture, what farmers want and need often differs from the simplifications of centrally planned priorities. Agricultural research and extension have, for example, a tendency to specialize on single commodities. But farmers' analysis will often specify a non-commodity need, such as multipurpose trees for agroforestry, or a rapidly vining legume to suppress weeds, or a range of vegetable seeds, or means to create, improve and exploit microenvironments, or technology for harvesting water, capturing and concentrating soil, or improving the supply of plant nutrients. As a result of past neglect, the potential for search and supply of such varied material and technologies seems still very large.

Search and supply have institutional implications. These include that

187

grass-roots extension staff and scientists have resources and are rewarded, for finding farmers' innovations and experiments and for stimulating and articulating realistic demand from farmers; and an ability of national and international agricultural research systems to respond with supplies of genetic material, principles and methods.

These reversals face two major obstacles. First, extensionists and scientists may not be rewarded for raising problems and making requests. Extensionists seen in the TOT and normal bureaucratic mode are there to pass on messages and packages downwards, not to multiply work for their senior officers by passing varied requests upwards. Second, most NARSs lack capacity to respond to needs and requests articulated by farmers for material or information. In practice, most management information systems are designed to feed information upwards to serve central management, rather than to draw it downwards to serve farmers. Six of the seven management information systems listed in 1987 for agricultural research in the Philippines were for central management; only one, the Research Information Storage and Retrieval System, was to provide information useful at the grass roots, and that was described in the future tense, with the statement that financial support was needed to extend it into the regions (Valmayor and Ramon, 1987). Many NARSs have poor institutional memories for research findings (see eg Kean and Singogo, 1988:48), and work often has to be repeated because earlier records cannot be found. Few, if any, are yet set up well enough to provide diverse information, genetic material and technologies to meet diverse local demand.

The practical implications are for agricultural research and extension organizations to make three changes: to encourage field staff to search for, support and spread farmers' innovations; to judge and reward staff by the requests they make upwards in response to analysis and demands by farmers; and to develop information and supply systems to respond to those demands.

(iii) Incentives. As with any new paradigm, professionals who innovate in the farmer-first mode risk being marginalized. In the short term, the safest route to promotion will often seem to be work on-station not on-farm; on irrigated agriculture, not rainfed (and least of all on unreliable rainfed); on a single commodity, not complex combinations; on industrial, commercial and major cereal crops not low status subsistence food crops; with quick maturing annuals not slow maturing perennials like shrubs and trees; and with validation through standard experimental design not farmers' adoption. Nor does improving complex, diverse and risk-prone (CDR) farming lend itself to the statistical testing methods taught in textbooks, involving as it often does complex and multiple simultaneous change, for example, agroforestry combined with water harvesting, growing fish with rainfed rice, home gardening with several canopies, or the creation and exploitation of protected microenvironments in semi-arid conditions. More papers can be produced more reliably by using conventional methods on conventional crops in conventional environments, where there is already a good information base, than by using unconventional methods on unconventional

agricultural practices in unconventional environments. Where promotions boards judge candidates only by adherence to standard methods, or numbers of publications, rather than farmers' adoption, then pioneers in farmer-first modes will not do as well as their less innovative colleagues.

The rapid transfer of agricultural research staff poses a further problem especially in sub-Saharan Africa. The costs in lost continuity and effectiveness in formal on-station research are well known. Less well recognized is the way in which rapid turnover reduces incentives for staff to build up relations with farmers, and undermines farmers' confidence in them.

The practical implications of these obstacles are to develop enabling conditions and incentives. The several forms these can take include the following:

- assessing research staff less on publications, and extension staff less on the achievement of targets; and both more on the demands and searches they initiate on behalf of farmers, on farmers' interest and innovation and on adoption and spread of technology;
- rewarding those who pioneer and write about new methods. Until recently, farmer-first research methods were not much the subject of articles in the harder scientific journals, but as the Summer 1988 issue of *Experimental Agriculture* (Farrington, 1988) has shown, this is changing. As scientists come to realize that they can publish articles about their methods and experiences, and that these bring national and international recognition, publishing disincentives should not just disappear but be reversed;
- ensuring more continuity for scientists in field posts. This may be difficult for many reasons. Fortunately, where lack of staff continuity is endemic, experimenting farmers and local organizations may be able, more and more, to provide their own continuity;
- networking between farmer-first researchers, providing mutual support and recognition. (For accessible opportunities see Appendix.)

The strongest incentive, though, is professional and personal satisfaction. Those who make reversals and changes in directions like those in this book, and who work collegially with farmers, soon find it intellectually and professionally exciting, enjoyable, and even fun, with the supreme reward of effectively helping farmers to do better. This is the most hopeful aspect. For even if other conditions are adverse, more and more will want to work in the farmer-first mode for the simple and sound reason that it satisfies and succeeds.

Methods and interactions

In themselves, these three things – decentralization and resources, organization for search and supply and providing incentives – are not enough. Much also depends on what is done and how it is done – on the methods available and the quality of interactions.

The need here is to develop further, describe and disseminate farmer-first methods in detail. Just as the aim is to widen choice of practices for

189

resource-poor farmers, so it is to widen choice of methods for scientists and extensionists. Some of these are methods for decentralization, for search and supply and for farmers' experiments; yet others are for interactions between professionals and farmers. Many such methods are now known. Those that are most promising deserve to be evaluated, written up and made accessible through manuals and practical training.

The more important methods to be developed and described include:

- aiding farmers' analysis and learning their agendas;
- getting started with families and communities;
- finding out about agricultural research (for NGOs);
- finding and supporting farmers' experiments;
- convening and assisting groups;
- convening and managing innovator workshops;
- searching, and supplying farmers with what they want and need;
- designing and managing incentives for scientists;
- communicating: farm family and outsider face-to-face.

This last, concerning the quality of interaction between farmers and scientists, is as crucial as it has been neglected. Most accounts and manuals concentrate on the mechanics of methods, as though rules guarantee results. This is not so. As social anthropologists, sociologists and some psychologists know, and as is only commonsense, the quality of the face-to-face relationship can make or mar an interview or discussion; and much depends on mutual respect and rapport.

Good advice is available (see eg Rhoades 1982; Grandstaff and Grandstaff 1987) but one may still ask how many scientists and extensionists have a grounding in the significance of non-verbal cues, of seating arrangements, of demeanour and manners and of that respect for and interest in people and what they have to show and say which makes for free and open communication.

Even good manuals and training for farmer-first methods and manners cannot by themselves guarantee good results. After institutions, incentives and interactions, there remains personality. Personal styles and aptitudes differ. The contrast between the closed blueprint approach to development and the open learning process (Korten 1980, 1984) parallels the contrast between TOT and farmer first. Some people are more at home with blueprints, with fixed plans and rules, and with clear ideas of what is expected and what will be officially rewarded. For them, the TOT mode fits better. Others are more at ease with learning processes, with open-ended exploration, with deciding for themselves how to proceed as they go along, and with the reward of knowing in themselves that they have done well. They will be better with the farmer-first mode.

A pluralist strategy

For farmer-first reversals, pluralism is one key to effective action. Individuals have different inclinations, aptitudes and opportunities, and these change over time. Organizations have different potentials, and these vary

190

between countries, regions and environments, and also change. There is no standard situation and no one formula, but there are questions of where to start.

Besides NARSs, the obvious natural leaders at first sight are the International Agricultural Research Centres. They are seen as prestigious sources of innovation, and they set standards for agricultural research. They train many of the more able national scientists. Their publications are easily available and widely consulted. They do, though, have disadvantages. At least one Centre (ICRISAT) has a mandate which is said to impede on-farm and with-farmer technology generation. The number of non-economist social scientists is everywhere low, and sometimes derisory. Many of the Centres' staff do not speak local vernaculars and so cannot listen directly to farmers. Excellent facilities, normal professional aspirations and high status frontiers such as biotechnology, combine to hold scientists at the central research stations and out of contact with farmers. To their credit, CIP (the International Potato Centre) in Peru and CIAT in Colombia have pioneered and popularized farmer-first methods and some staff at IRRI in the Philippines are active. But the numbers of staff involved are still small, and it remains to be seen how far and fast they and others can go. For the present, the powerful influences of the international Centres mostly reinforce the conventional TOT paradigm. The Centres are still more of the problem than of the solution. But they need not remain so. With a new vision and understanding, they could lead in developing, improving and spreading the farmer-first approach and methods.

Agricultural universities and faculties, and management institutes which train scientists and extensionists, are another focus for change. Some universities are bastions of conservatism, doggedly reproducing narrow professionalism in their students. Others are more open and innovative. By changing their curricula and teaching, by rewriting their textbooks and by introducing learning from and with farmers, universities and training institutes could help mould and transform the values and behaviour of new generations of scientists and extensionists.

Given their influence, size and coverage, these large organizations – International Agricultural Research Centres, NARSs, national agricultural extension organizations and universities and faculties – must in the longer term be transformed if the gross imbalance between TOT and farmer-first is to be corrected. To achieve this on their own, in isolation, would be difficult though. Fortunately, three other, smaller-scale, types of organization and arrangement provide more favourable environments for reversals and change. These are projects, NGOs and farmers' organizations.

Special projects, working in various combinations with NARSs, are well represented by the contributions to this book: the Agricultural Research Planning Teams in Zambia (Kean), the Agricultural Technology Improvement Project in Botswana (Norman et al), the Tropsoils Project in West Sumatra (Colfer et al), the Agricultural Research and Production Project in Nepal (Mathema and Galt) and the Farming Systems Development Project in the Eastern Visayas in the Philippines (Lightfoot et al, Repulda et al).

These projects combined special resources with staff who wished to work closely with farmers, and who had the freedom to do so.

For their part, international and national NGOs have advantages. It is true that they are scattered, of variable quality and usually small. They have also, as a whole, tended to be weak on the technical side of agriculture and inexpert at making links with formal agricultural research. Change, though, is rapid. In the late 1980s, many have been shifting their priorities, staff recruitment and training towards agriculture. NGOs have a comparative advantage, especially when they can maintain the same good staff in the field in the same place for a number of years. Some, like World Neighbors, the Central Mennonites Committee, the Aga Khan Rural Support Programmes, OXFAM, and Save the Childen Fund, already have a track record in farmer-first innovation. NGOs like these find it easier than large bureaucracies to avoid the trap of TOT, to recruit and maintain sensitive staff in the field, to be close to farmers, to encourage their participation and to act in farmer-first roles.

Farmers' organizations are a form of national NGO of growing significance. They have an increasing capacity to make demands on NARSs and to influence and sometimes even fund research. They tend, though, to represent the better-endowed farmers and those who produce for standard large-scale markets. The resource-poor farmers of CDR agriculture tend to be unorganized and to have diverse needs which defy simple aggregation. For them, demand-pull will always be weaker and the responsibility for putting their priorities first rests much more with other NGOs and with individual professionals.

A plurality of organizations can combine to gain strength in diversity. This has been observed in Eastern Bolivia (Thiele et al 1988), where area-based development projects, NGOs and producers' organizations have been 'intermediate users' of technology, exercising demands on the formal research organization on behalf of farmers. There and elsewhere, both organizational links and staff careers are becoming more varied and plural. Projects provide resources for scientists to travel and for fieldwork with farmers. NGOs arrange visits for farmers to other areas. NGOs overcome their lack of agricultural competence by recruiting staff who leave government service, or, as in the Sudan, by paying them supplements while they remain on the government payroll.

A pluralist strategy, involving a variety of large and small organizations, partly answers questions of cost-effectiveness in the use of scientists' time. Sometimes the opportunity costs of scientists working on CDR agriculture may appear high, for instance if an African country has only a few scientists to work on an industrial crop of national importance. Further, there are usually far fewer scientists per farming system in CDR than in green revolution agriculture. A case could be put that scientists' impact working on CDR agriculture will be low. This has been evident so far. Concern is expressed that so much of the output of research is not adopted by farmers, with the rate of rejection in India informally estimated at 80 per cent or more, with probably a higher figure for rainfed agriculture.

The farmer-first mode promises greater cost-effectiveness. Where NGOs or extension agents are the convenors, catalysts and communicators, scientists can be used sparingly as consultants. When farmers play a full part, they themselves take account of local diversity in a manner that makes low demands on scarce scientific staff. When scientists spend more time searching for genetic material, technologies and principles for farmers to try, adapt and choose between, they may have more impact. Above all, putting farmers' agendas first and helping them to meet their priorities should be a sure path to good use of time. In a plural farmer-first approach, farmers, NGO workers, extensionists and agricultural researchers can specialize and support each other, with farmers and their groups and networks doing most, and the others serving them. In making the most of scarce staff, pluralism should pay off.

Practical action: starting and sustaining change

Professionals concerned with agricultural innovation, research and extension – whether they are farmers, or physical, biological or social scientists, and whether they are independent or working in universities, training institutes, government departments or NGOs – will have found in this book many ideas for what they might do. Non-farming agricultural professionals, just like resource-poor farmers, are faced with diversity and complexity, and similarly need a repertoire of methods so that they can be versatile and adaptable. Yet more methods are being generated, and several regular sources of information about them which can be obtained free for the asking are listed in the appendix.

At a personal level, it is tempting to say that nothing can be done until a whole bureaucratic and professional system changes. Usually, though, there is room for manoeuvre. Some steps can be taken; a start can almost always be made. Even if the start is small and progress slow, it may be the seed of a self-sustaining movement. In the spirit of the learning process approach to development, it is better to start, to do something and to learn on the way, than to wait for better conditions before acting.

In the spirit of pluralism, action can and should start in many places. But not everything can be done at once. There are questions of how and where to start.

Two principles help here. The first is to start where it is easier, simpler and quicker, while weighing the danger of biases against poorer farmers. It is better to start and learn by doing and through mistakes than to wait for perfect conditions. By starting, experience is gained and confidence built up.

The second principle is to change behaviour before attitudes. Preaching about attitudes invites acquiescence without deep change. Action means experience gained and that, more than exhortation, reorients attitudes and habits of thought.

Taking these two principles together, analysis by and with farmers appears the most promising point of entry, followed by search, choice and experiment. A basic question to ask is what farmers would like in their basket of choices. From this question follow demands which reverse the

193

normal top-down flow. Whether a department of agriculture, a university, an NGO or combinations of these can handle such requests can then be put to the test. Activities and roles then have to change. Procedures to accept and handle demands are required. Information systems for management from below have to be created and made to work. Subsequently, other elements of the paradigm become active, with testing and experiments by farmers and consultative support by others.

It is one thing to start and establish a bridgehead. It is quite another to sustain and spread it. The experiences reported by ISNAR's On-Farm Client-Oriented Research project in nine national institutes are sobering. They include difficulty maintaining an interdisciplinary focus, vulnerability to the withdrawal of special support, a tendency to methodological stagnation, a loss of early enthusiasm and of farmer participation, and a career ladder which leads away from collaboration on-farm and towards specialization on-station (Merrill-Sands 1988; Ewell 1988; von der Osten et al 1988). With farmer-first, similar problems can be expected but also differences. The approach and methods described by the contributors to this book go further than most on-farm research, exploiting as they do the comparative advantage of farmers' knowledge, continuity and capacity for innovation. When it is farmers, with their full experience of their own farming systems, who analyse, experiment, monitor and make judgments, it is less important to sustain an interdisciplinary focus; farmers' enthusiasm and participation are more likely; and if outside support weakens, farmers can carry on on their own, and make their own demands on the research system, strengthened by their personal interest and participation. Compared with on-farm research in the TOT mode, the farmer-first mode promises to be feasible with a lighter touch and sustainable with less outside support.

Finally, for professionals to innovate by working in the farmer-first mode demands vision and leadership on the part of those with power and responsibility. These include senior officials in capital cities, vice-chancellors and deans, directors of research stations, leaders of teams and senior staff in regional, provincial and district headquarters, as well as in aid agencies and NGOs. Leaders can act like normal professionals and normal bureaucrats who simplify, standardize and stifle; or they can break out and encourage and support initiative and change, providing resources and room for manoeuvre for those under their management who have the aptitude and will to work in new participatory ways; and they can reorganize departments, procedures and management information systems so that searches can be made to meet farmers' demands and fill their basket with choices.

Alliances and mutual support also help. Those who see or sense the potential will do well to seek out and support like-minded fellow professionals in their own and other organizations. Shared ideas and experiences speed up learning. If those in this book provide stimulus and encouragement, they will have served their purpose. And if the new paradigm fulfills its promise, and is accepted and practiced much more in the 1990s and the 21st century, then those who take risks now to support, develop and spread it will not have acted in vain.

For the stakes are high. Over a billion people are supported by the third agriculture. The challenge is to enable many of the poorer among them to secure better and more sustainable livelihoods from their complex, diverse and risk-prone farming when normal agricultural research has so largely failed. This book points to new potentials. It shows that reversals in the farmer-first mode can be effective for farmers and exciting for professionals. A quiet revolution has already started, but it is scattered and still small-scale. Which countries, institutions and individuals will now lead remains to be seen. Change depends on personal decisions and action. Those who now explore the frontiers of farmer participation cannot expect Nobel prizes, or be confident of early recognition or promotion; but they will be joining a vanguard. Their rewards, more surely, will be the exhilaration of pioneering, the satisfaction of seeing innovations spread and the knowledge that through their work, poor farm families are being truly served.

Notes

[1] Based on discussions in the ITK study group and informal comments by Anil Gupta, Roland Bunch, Lori Ann Thrupp and Ed Barrow.
[2] 30,000 scientists in the whole of India's agricultural research system, adding universities and private sector to ICAR's 6,000.
[3] It is perhaps worth noting that research project management hardly exists as yet in science in western countries, and almost nobody has been trained in it. This may explain some of the confusion in British and American science policy, as well as the problems which arise as financial constraints are tightened.

Appendix: Sources of further information which are free to Third World readers

Some of the main sources of information are still highly priced journals like *Agricultural Administration and Extension* (now merged with *Agricultural Systems* to become the *Journal of Agricultural Systems*) which many libraries let alone individuals, cannot afford. Access to free information is now harder because of the astonishing error of judgment of FAO in suspending or terminating *Ceres* and *Unasylva*, which were free to Third World Readers. Readers who wish to remain up-to-date with developments and/or to share their methods and experiences with others do have other sources available free of cost. In some cases, there are charges for those on first world salaries.

CIAT – Centro Internacional de Agricultura Tropical, AA 6713, Cali, Colombia. A bibliography on farmer participatory research, with 35 per cent of the entries in Spanish. Also a video on farmer participation in agricultural research can be obtained free in VHS or Beta for USA standard television by sending a blank tape to the Co-ordinator, Participatory Research Projects, CIAT.

CIKARD News, concentrating on indigenous knowledge systems, decision-making, organization and innovations, available from Center for Indigenous Knowledge for Agriculture and Rural Development, Iowa State University, 319 Curtiss Hall, Ames, IA 50011, USA.

CTA – Technical Centre for Agricultural and Rural Co-operation (ACP–EEC Lome Convention), Postbus 380, 6700 AJ Wageningen, The Netherlands. Established to improve access to technical information on agricultural development for ACP (Africa, Caribbean, Pacific) states. ACP nationals may request free subscriptions to the bimonthly bulletin *Spore* and free access to Question-Answer and Document Delivery Services.

Agroforestry Today. Available from the International Council for Research in Agroforestry, PO Box 30677, Nairobi, Kenya, quarterly.

ILEIA Newsletter. Individuals and organizations in the Third World may request free subscriptions from Information Centre for Low External Input Agriculture, PO Box 64, 3830 AB Leusden, The Netherlands, quarterly, 28pp. Two issues in May 1988 covered sources of information: *Towards Sustainable Agriculture: Part 1, abstracts, periodicals, organizations*; *Part 2 bibliography*. The October 1988 issue was on *Participative Technology Development* (also in French). Also free to the Third World, *Proceedings of the ILEIA Workshop on Participatory Technology Development in Sustainable Agriculture, April 1988*.

International Institute for Environment and Development, 3 Endsleigh Street, London WC1H 0DD, UK, produces an informal newsheet, *RRA Notes*, with information on rapid rural appraisal methods and experiences. Also McCracken, JA., Pretty JN and Conway GR, 1988, *An Introduction to Rapid Rural Appraisal for Agricultural Development* is available free to the Third World. The Drylands Development Programme, IIED, publishes *Haramata*, a quarterly newsletter, which aims to network between researchers, NGOs and policy makers involved in dryland development questions, and investigates current dryland development issues. Available free to the Third World.

Overseas Development Institute, Regent's College, Inner Circle, Regent's Park, London NW1 4NS, UK, has an *Agricultural Administration (Research and Extension) Network Newsletter and Discussion Papers*, including farmer participatory research, issued twice yearly. Also available while stocks last: *Experimental Agriculture* vol 24 part 3 1988 (special issue on farmer participatory research) and John Farrington and Adrienne Martin, Farmer Participatory Research: a review of concepts and Practices, *ODI Occasional Paper No 9*. A Network Paper containing abstracts of some 200 papers on farmer participatory research will be available in spring 1989. ODI has a specialist Agricultural Administration library with computerized search facilities available to visitors. ODI's other networks (on Social Forestry, Pastoral Development and Irrigation Management) operate similarly and also carry papers on participatory approaches to management and technology development.

Reading Rural Development Communications Bulletin, published by the Agricultural Extension and Rural Development Department, University of Reading, London Road, Reading RG1 5AQ, UK, twice yearly, covering a range of participatory approaches. Free of charge on an exchange basis, or £7.50 per four issues.

World Neighbors in Action, a newsletter with practical new ideas and tested practices in Third World agriculture, published by World Neighbors 5116 Portland Avenue, Oklahoma City, OK 73112, USA. Available free to the Third World.

Abbreviations

AF	Agroforestry
ARPP	Agricultural Research and Production Project (Nepal)
ARS	Agricultural Research Service (India)
ATIP	Agricultural Technology Improvement Project (Private Bag 2427, Gaborone, Botswana)
BARI	Bangladesh Agricultural Research Institute (Joydebpur, Gazipur, Bangladesh)
BAU	Bangladesh Agricultural University
BFD	Bureau of Forest Development (Philippines)
bn	billion
CGIAR	Consultative Group on International Agricultural Research (CGIAR Secretariat, 1818 H Street NW, Washington DC 20433, USA)
CIAT	Centro Internacional de Agricultura Tropical (International Centre for Tropical Agriculture) (CIAT, AA 6713, Cali, Colombia)
CIMMYT	Centro Internacional para Mejoramiento de Maiz y Trigo (International Centre for the Improvement of Corn and Wheat) (Londres 40, AP 6–641, Mexico 06600 DF)
CIP	Centro Internacional de la Papa (International Potato Centre) (Apartado 5969, Lima, Peru)
CTA	Technical Centre for Agricultural and Rural Co-operation (Postbus 380, 6700 AJ Wageningen, The Netherlands)
FF	Farmer First
FPR	Farmer Participatory Research
FSR	Farming Systems Research
IARC(s)	Intenational Agricultural Research Centre(s)
ICAR	Indian Council for Agricultural Research
ICP	Integrated Cereals Project (Nepal)
ICRAF	International Council for Research on Agroforestry (PO Box 30677, Nairobi, Kenya)

ICRISAT	International Crops Research Institute for the Semi-Arid Tropics (Patancheru PO, Hyderabad, AP 502 324, India)
IDS	Institute of Development Studies (University of Sussex, Brighton, East Sussex, BN1 9RE, UK)
IIED	International Institute for Environment and Development (3 Endsleigh Street, London, WC1H 0DD UK)
IITA	International Institute for Tropical Agriculture (PMB 5320, Ibadan, Nigeria)
ILCA	International Livestock Centre for Africa (PO Box 5689, Addis Ababa, Ethiopia)
ILEIA	Information Centre for Low External Input Agriculture (c/o ETC, PO Box 64, 3830 AB Leusden, The Netherlands)
IRRI	International Rice Research Institute (PO Box 933, Manila, Philippines)
ISNAR	International Service for National Agricultural Research (PO Box 93375, 2509 AJ The Hague, The Netherlands)
ITK	Indigenous technical knowledge
km	kilometre
m	metre, million
NARP	National Agricultural Research Project (India)
NARS(s)	National Agricultural Research System(s)
NGO(s)	Non-Government Organization(s)
ODI	Overseas Development Institute (Regent's College, Inner Circle, Regent's Park, London NW1 4NS)
RRA	Rapid Rural Appraisal
RPF	Resource poor farmer
R&D	Research and Development
TOT	Transfer of Technology
TSP	Triple Superphosphate
UDP	Upland Development Programme (Philippines)
USAID	United States Agency for International Development
UPLB	University of the Philippines, Los Banos
WN	World Neighbors (5116 Portland Avenue, Oklahoma City, OK 73112, USA)

References and sources

An asterisk (*) indicates a paper presented at the IDS Workshop on Farmers and Agricultural Research: Complementary Methods, held at the Institute of Development Studies, University of Sussex, UK 26–31 July 1987. Only the short title of such papers is given, together with a reference to any published version. Where papers have not been formally published, copies can be obtained from: Institute of Development Studies, University of Sussex, Brighton, BN1 9RE, England, or from Overseas Development Institute, Regent's College, Inner Circle, Regent's Park, London, NW1 4NS

Abedin, MZ, 1982, 'Proceedings of the workshop on experience gained by progressive farmers in potato cultivation in Rangpur District', E&R Project, Bangladesh Agricultural Research Institute, Ishurdi, Bangladesh

*Abedin, MZ, and Haque, MF, 1987, 'Learning from farmer innovations and innovator workshops', IDS Workshop. For shortened version, see section 3.5 above

African Development and Economic Consultants (ADEC), 1983, Agroforestry systems evaluation survey, ADEC, Nairobi

Aguila, F Jr, 1982, 'Social forestry for upland development: lessons from four case studies', Institute of Philippine Culture

Alao, J, 1980, 'Understanding small farmer adoption behaviour: the Nigerian experience', in *Agurla Lecture Series*, 44, University of Ife Press, Ife-Ife, Nigeria

Altieri, MA, 1983, 'Agroecology: the scientific basis of alternative agriculture', Division of Biological Control, University of California, Berkeley, CA

Amir, P, and Knipscheer, HC, 1987, 'Application of the environment-behaviour-performance model in farming systems research', *Agricultural Administration and Extension*, 25, pp 161–176

Ashby, JA, 1986, 'Methodology for the participation of small farmers in the design of on-farm trials', *Agricultural Administration and Extension*, 22, pp 1–19

Ashby, JA, 1987, 'The effects of different types of farmer participation on the management of on-farm trials', *Agricultural Administration and Extension*, 24, pp 235–252

*Ashby, JA, Quiros, CA, and Rivera, YM, 1987, 'Farmer participation in on-farm varietal trials', IDS Workshop. For shortened version, see sections 3.2 and 3.4 above. Available as ODI *Agricultural Administration (Research and Extension) Network Discussion Paper 22*, December 1987

Atta-Krah, AN, and Francis PA, 1987, 'The role of on-farm trials in the evaluation of composite technologies: alley farming in Southern Nigeria', *Agricultural Systems*, 23, pp 133–152

*Baker, G, Knipscheer, HC, Neto, Jose de Souza, 1988, 'The impact of regular research field hearings in on-farm trials in northeast Brazil', *Experimental Agriculture* Vol 4, part 3, pp 281–288

Bari, F, 1974, 'An innovator in a traditional environment', Bangladesh Academy for Rural Development, Kotbari, Comilla, Bangladesh, 55 pp.

Barrow, EGC, 1985, 'An analysis of human and environmental factors in the agricultural development of East Pokot', Nginyang Division, Baringo District, Kenya. Master's thesis, Antioch University, USA

*Barrow, EGC, 1987, 'Extension and learning examples from the Pokot and Turkana', IDS Workshop

Basant, R, 1988, 'The diffusion of agro-mechanical technology for Indian rainfed farming: an exploratory analysis', *Agricultural Administration (Research and Extension) Networking) Paper* 24 ODI, London

Bellman, BL, 1984, *The Language of Secrecy: symbols and metaphors in Poro Ritual*, New Brunswick: Rutgers University Press

Biggs, SD, 1978, 'On-farm and village level research: an approach to the development of agricultural and rural technologies', in *Economic Problems in Transfer of Agricultural Technology. Proceedings of a National Seminar*, Indian Agricultural Research Institute, New Delhi

Biggs, SD, 1979, 'Timely analysis in programmes to generate agricultural technologies', in Conference on Rapid Rural Appraisal, Institute of Development Studies, University of Sussex, Brighton

Biggs, SD, 1980, 'Informal R and D', *Ceres*, 13(4), pp 23–6

Biggs, SD, 1981, 'Sources of innovation in agricultural technology', *World Development*, Vol 9, no 4, pp 321–336

Biggs, SD, 1983, 'Institutions and decision-making in agricultural research', in *The Economics of New Technology in Developing Countries*, Frances Pinter, London

Biggs, SD, 1987, 'Interactions between resource-poor farmers and scientists in agricultural research', School of Development Studies, University of East Anglia, Discussion Paper for OFCOR research group; ISNAR study on Organization and Management of On-farm Research

Biggs, SD, 1988, *Resource-poor Farmer Participation in Research: a synthesis of experiences from nine national agricultural research systems*, ISNAR, The Hague

Biggs, SD, and Farrington, J, 1989, 'Social science analysis in agricultural research: a review and conceptual framework', IDRC, Ottawa

Box, L, 1982, 'Food, feed or fuel? Agricultural development alternatives and the case for technological innovation in cassava cultivation', *Quarterly Journal of International Agriculture*, Special Issue: 34–48

*Box, L, 1987a, 'Experimenting cultivators', IDS Workshop. For shortened version, see section 2.2 above. Available in full as 'Experimenting cultivators: a methodology for adaptive agricultural research', ODI Agricultural Administration (Research and Extension) Network *Discussion Paper* 23, December 1987

Box, L, 1987b, 'Knowledge, networks and cultivators: cassava in the Dominican Republic', International Course for Rural Extension, Wageningen

Box, L, and Doorman, FJ, 1985, 'The adaptive farmer: sociological contributions to adaptive agricultural research on cassava and rice cultivation in the Dominican Republic', Department of Rural Sociology of the Tropics and Sub-tropics, Agricultural University, Wageningen

Braidwood, RJ, 1967, *Prehistoric Men*, Scott, Foresman and Co, Glenview, Illinois

Brammer, H, 1980, 'Some innovations don't wait for experts: a report on applied research by Bangladesh peasants', *Ceres*, 13 (2), pp 24–28

Brammer, H, 1982, 'Crop intensification: why and how lessons from peasants in Bangladesh', *Ceres*, 15(3), pp 43–45

Brokensha, D, Warren, D, and Werner, O, (eds.) 1980, *Indigenous Knowledge Systems and Development*, University Press of America, Lanham, Maryland

Buck, L, 1988, *Agroforestry Extension Training Source Book*, CARE International, New York, 540 pp

Bunch, R, 1985, *Two Ears of Corn: a guide to people-centered agricultural improvement*, World Neighbors, 5116 North Portland, Oklahoma City, Oklahoma 73112

*Bunch, R, 1987, 'Small farmer research', IDS Workshop. For shortened version, see section 2.1 above

Byerlee, DK, and Collinson, MP, 1980, *Planning Technologies Appropriate to Farmers: concepts and procedures*, CIMMYT, Mexico

Caldwell, JS, and Lightfoot, C, 1987, 'A network for methods of farmer-led systems experimentation', *FSSP Newsletter*, 5(4): pp 18–24

Carlier, 1987, *Understanding Traditional Agriculture: Bibliography for Development Workers*, ILEIA, Leusden, The Netherlands

Carson, B, 1987, 'Appraisal of rural resources using aerial photography: an example from a remote hill region in Nepal', in KKU 1987, *Proceedings of the 1985 Conference on Rapid Rural Appraisal*, pp 174–190

Casley, D, and Kumar, K, 1987, *Project Monitoring and Evaluation in Agriculture*, The Johns Hopkins University Press

Chambers, R, 1983, *Rural Development: putting the last first*, Longman, Harlow

Chambers, R, 1988, 'Direct Matrix Ranking in Kenya and West Bengal', *RRA Notes 1*, IIED, London, pp 13–18

Chambers, R, forthcoming, 'Farmer-first: a practical paradigm for the third agriculture'. In M Altieri and S Hecht (eds) *Agroecology and Small Farm Development*, CRC Press, Florida

Chambers, R, and Ghildyal, BP, 1985, 'Agricultural research for resource-poor farmers: the farmer-first-and-last model', *Agricultural Administration and Extension*, 20, pp 1–30

Chambers, R, and Jiggins, J, 1986, 'Agricultural Research for resource-poor farmers: a parsimonious paradigm', *Discussion Paper* 220, IDS, University of Sussex

Chambers, R, and Longhurst, R, 1986, 'Trees, seasons and the poor', *IDS Bulletin*, 17 (3)

Chambers, R, and Leach, M, 1987, 'Trees to meet contingencies: savings and security for the rural poor', *Discussion Paper* 228, Institute of Development Studies, University of Sussex, 22 pp

Chapman, B, 1984, Diet and production survey memoranda 1, 2, 3, Tropsoils Research Report, Sitiung, Indonesia

*Charoenwatana, T, 1987, 'Farmers and agricultural science', IDS Workshop

Chavangi, NA, 1988, 'Problem definition and a statement on the case for an expanded awareness programme in Kakamega district', Kenya Woodfuel Development Project

*Chavangi, N, and Ngugi, A, 1987, 'Innovatory participation in programme design: tree-planting for increased fuelwood supply for rural households in Kenya', IDS Workshop

*Chaves, LE, 1987, 'Some experiences of farmers' participation in Colombia', IDS Workshop

Colfer, CJP, 1983, 'On communication among "unequals"', *International Journal of Intercultural Communication*, 7, pp 263–283

*Colfer, CJP, 1987a, 'Intra-team collaboration in the Tropsoils project', IDS Workshop

*Colfer, CJP, 1987b, 'On farmer-researcher interaction', IDS Workshop

*Colfer, CJP, 1987c, 'Farmer involvement in the Tropsoils project: two complementary approaches', IDS Workshop. See section 3.8 for a shortened version

*Colfer, CJP, 1987d, 'How to ascertain which crops are grown', IDS Workshop

*Colfer, CJP, Evensen, C, Evensen, S, Fahmuddin Agus, D, Gill, D, Wade, A and Chapman, B, 1985, '"Transmigrants" gardens: a neglected research opportunity', Proceedings, Centre for Soil Research, Annual Technical Meetings, Bogor, Indonesia

*Colfer, CJP, Gill, D, and Fahmuddin Agus, D, *Indigenous Agricultural Models: a source of scientific insight*, forthcoming

Collinson, MP, 1981, 'A low-cost approach to understanding small farmers', *Agricultural Administration and Extension*, 8, pp 433–50

Collinson, MP, 1982, 'Farming systems research in Eastern Africa the experience of CIMMYT and some national agricultural research services, 1976–81', *International Development Paper No. 3*, Michigan State University, East Lansing

Collinson, MP, 1987, 'Farming systems research: procedures for technology development', *Experimental Agriculture*, Vol 23, pp 365–86

Collinson, MP, 1988, 'The development of African farming systems: some personal views', *Agricultural Administration and Extension*, Vol 29, No 1, pp 7–22

Conroy, C, and Litvinoff, M, eds, 1988, *The Greening of Aid: Sustainable Livelihoods in Practice*, Earthscan Publications, London

Conway, GR, 1985, 'Agroecosystem Analysis', *Agricultural Administration*, Vol 20, pp 31–55

Conway, GR, 1986, 'Agroecosystem analysis for research and development', Winrock International, Bangkok

Conway, GR, 1987a, 'The properties of agroecosystems', *Agricultural Systems*, 24, pp 95–117

*Conway, GR, 1987b, 'Diagrams for farmers', IDS Workshop. For shortened version, see section 2.5 above

Conway, GR, and Sajise, PE, 1986, 'The agroecosystems of Buhi', Programme on Environmental Science and Management, University of Philippines, Los Banos

Conway, GR, McCracken, JA, and Pretty, JN, 1987, *Training Notes for Agroecosystem Analysis and Rapid Rural Appraisal*, second edition, Sustainable Agriculture Programme, International Institute for Environment and Development, London

De Jager, A, 1988, 'Towards self-experimenting village groups', paper for workshop on Operational Approaches for Participative Technology in Sustainable Agriculture, ILEIA, Leusden, 11–12 April 1988

Dharampal, 1971, *Indian Science and Technology in the Eighteenth Century*, Impex India, Delhi

*Edwards, RJA, 1987a, 'Farmers' knowledge: utilization of farmers' soil and land classification', IDS Workshop

*Edwards, RJA, 1987b, 'Farmers' groups and panels', IDS Workshop

*Edwards, RJA, 1987c, 'Mapping and informal experimentation by farmers: agronomic monitoring of farmers' cropping systems', IDS Workshop

*Eklund, P, 1987, 'Low-cost diagnostic methods for low-input strategies in sub-Saharan Africa', IDS Workshop

ERC, 1988, *Rapid Rural Appraisal: a closer look at rural life in Wollo*, Ethiopian Red Cross Society, Addis Ababa and the International Institute for Environment and Development, London

Evensen, C, Yost, R, and Wade, M, 1985, 'Source and management of green manure – preliminary uniformity trial', *Field Research Brief No. 16*, Tropsoils Project, Sitiung, Indonesia

Fahmuddin Agus, D, Wade, MK and Prawirasumantri, J, 1985, 'Effects of post-clearing methods on soil properties and crop production', *Field Research Brief Nos. 7, 18 and 25*, Tropsoils Project, Sitiung, West Sumatra, Indonesia

Farrington, J, 1988, (ed), *Experimental Agriculture*, Vol 24, part 3, with 'Farmer Participatory Research: Editorial Introduction', pp 269–279

Farrington, J, 1989, 'Farmer participation in agricultural research', *Food Policy*, Vol 14, No 2

Farrington, J and Martin, A, 1988, 'Farmer Participation in Agricultural Research: a review of concepts and practices', Agricultural Administration Occasional Paper No 9, ODI, London

Feldstein, H, Poats, S, and Rocheleau, D, 1987, 'Intra-household analysis and on-farm research and experimentation'. Paper presented at the CIMMYT Networkshop on Intra-Household Dynamics and Farming Systems Research, Lusaka, Zambia, 25–30 April

Feldstein, H, and Jiggins, J, forthcoming, *Methodologies Handbook: Intra-household Dynamics and Farming Systems Research and Extension*, Population Council, New York

Fernandez, ME, 1988, 'Towards a participatory approach: new demands on researchers and research methodologies', *ILEIA Newsletter* Vol 4, No 3, pp 15–17

*Fernandez, ME and Salvatierra, H, 1987, 'Design and implementation of participatory technology validation in highland communities of Peru'. See section 3.7 for a shortened version of this paper; see also Farming Systems Research Symposium, Kansas State University, Manhattan, Kansas, 5–8 October 1986

Flint, M, 1986, 'IFPP Farm Management Survey', mimeo, Botswana

Floquet, A, 1989, 'Conservation of soil fertility by peasant farmers in Atlantic Province, Benin', in J Kotschi (ed) *Ecofarming Practices for Tropical Smallholdings – Research and Development in Technical Cooperation*, GTZ, Eschborn

Fresco, LO, 1986, 'Cassava in shifting cultivation: a systems approach to agricultural technology development in Africa', Royal Tropical Institute, Amsterdam

Freudenberger, C Dean, 1988, 'The agricultural agenda for the Twenty First Century', *Kidma*, Israel Journal of Development, No 38, Vol 10, No 2, pp 32–36

Friesen, G, et al, 1982, 'Decade of dryland research in India', All India Coordinated Research Project for Dryland Agriculture, Santoshnagar, Hyderabad 500659, India

Friesen, GH, Venkateswarlu, J, Rastogi, BK, Subramaniam, V. Bala, 1982, 'A decade of dryland agricultural research in India 1971–1986', All-India Coordinated Research Project for Dryland Agriculture, Saidabad, Hyderabad 500659

FSRDD, 1986, 'Naldung Farming Systems Site "Samuhik Bhraman"', *Report No. 5*, Ministry of Agriculture, FSRD Division, Khumaltar, Nepal

FSRDD, 1987, 'Kotjahari Samuhik Bhraman and Proposed Research Program', *Report No. 7*, Department of Agriculture, FSRD Division, Khumaltar, Nepal

Fujisaka, S, 1988a, 'Farmer participation in upland soil conservation research and technology dissemination in the Philippines', Agricultural Economics Department, IRRI, Manila, Philippines

Fujisaka, S, 1988b, 'A method for farmer-participatory research and technology transfer: upland soil conservation in the Philippines', Agricultural Economics Department, IRRI, Manila, Philippines

Fussell, GE, 1965, *Farming Techniques from Prehistoric to Modern Times*, Pergamon Press, Oxford, UK

*Galt, D, 1987, 'Informal observations on institutionalizing FSR in Nepal', IDS Workshop

Ganapin, D, 1979, 'Factors of underdevelopment in Kaingin communities', College of Forestry, University of Philippines, Los Banos

Getahun, A, Ndungu, J, Kedir, R and Bashir, J, 1986, 'Agroforestry experimental designs used by Kenya Renewable Energy Development Project', Ministry of Energy and Regional Development, Nairobi

*Ghildyal, BP, 1987, 'Drought-prone rice environment', 'Farmers' evaluation of rice-breeding materials', 'Appropriate technology and farmer-to-farmer extension in East India' (three short notes), IDS Workshop

Giddens, A, 1979, *Central Problems in Social Theory*, Macmillan, Basingstoke, UK

Gilbert, EH, Norman, DW, and Winch, FE, 1980, *Farming Systems Research: a critical appraisal*, MSU Rural Development Paper No 6, Department of Agricultural Economics, Michigan State University, East Lansing, Michigan 48824

Gill, D, Kasno, A, and Adiningsih, JS, 1985, 'Response of Soybeans/Upland Rice and Soil K levels to K Fertilization and Green Manure Applications at Sitiung', *Field Research Briefs, Nos 8 and 13*, Tropsoils Project, Sitiung, West Sumatra, Indonesia

Gomez, K, 1977, 'On-farm assessment of yield constraints: methodological problems', in *Constraints to high yields on rice farms: an intensive report*, IRRI, Philippines

Goodell, G, 1982, 'Communication from farmer to scientist', (unpublished ms cited by Rhoades, 1987)

Grandstaff, SW and Grandstaff, TB, 1987, 'Semi-structured interviewing by multi-disciplinary teams in RRA', in KKU 1987, *Proceedings of the 1985 Conference on Rapid Rural Appraisal*, pp 129–143

Grigg, D, 1982, *The Dynamics of Agricultural Change: the historical experience*, Hutchinson, London

Groosman, A, 1986, 'Technology development and the improved seed industry in North–South perspective', Development Research Institute, Tilburg University, Netherlands

Gubbels, P, 1988, 'Peasant farmer agricultural self-development: the World Neighbors experience in West Africa', *ILEIA Newsletter*, Vol 4, No 3, pp 11–14

*Gupta, AK, 1987a 'Organizing the poor client responsive research system: can tail wag the dog?', IDS Workshop. See sections 1.4 and 2.6 above for extracts

*Gupta,AK, 1987b, 'Scientific perception of farmers' innovations', IDS Workshop. See section 1.4 above for extracts

Gupta, AK, 1987c, 'Technology for dry-farming: how the scientists, students and farmers view the challenge?' Indian Institute of Management, Ahmedabad

Gupta, AK, Patel, NT and Shah, RN, 1987, 'Matching farmers' objectives with technologists' objectives in dry farming regions', Centre for Management in Agriculture, Indian Institute of Management, Ahmedabad, mimeo

Haque, F, 1985, 'Proceedings of innovative farmers' workshop, 19–20 February, 1985', OFRD, Bangladesh Agricultural Research Institute, Jessore (mimeo, in Bangla)

Harrell-Bond B, 1986, Imposing Aid: emergency assistance to refugees, Oxford University Press, Oxford, New York, Nairobi

Hart, R, 1980, 'Agroecosistemas: conceptos basicos', CATIE, Turri-alba, Costa Rica

Hart, R, 1986, 'Research and development strategies to improve integrated crop, livestock and tree systems'. Paper presented at the International Agricultural Research Centers (IARC) workshop on Farming Systems Research, ICRISAT, Hyderabad, India, 17–21 February

Harwood, RR, 1979, Small Farm Development: understanding and improving farming systems in the humid tropics, Westview Press, Boulder, Colorado

Hatch, JK, 1976, The Corn Farmers of Motupe: a study of traditional farming practices in northern coastal Peru, Land Tenure Centre, Monograph No 1, Madison, University of Wisconsin

Haverkort, B, and Engel, P, 1985, 'The systems approach, agricultural development and extension'. IAC, Knowledge systems in agricultural development, Manual Workshop III, International Course on rural extension, International Agricultural Centre, Wageningen, The Netherlands

Haverkort, B, 1987, 'Agricultural Development and Agricultural Knowledge. In an International Perspective', paper presented at the 35th International Course on Rural Extension, Wageningen, The Netherlands, 1987, pp 36

Haverkort, B, 1988, 'Agricultural Production Potentials: inherent, or result of investment in technology development? The influence of technology gaps on the assessment of production potentials in developing countries'. ETC Foundation, PO Box 64, 3830 AB Leusden, The Netherlands

Hendry, P, 1987, 'Research on farming systems offers new perspectives', Ceres, Vol 20, No 6, November–December, pp 13–15

Hildebrand, PE, 1979a, 'Initial characterization – the rapid survey or Sondeo', ICTA, Guatemala

Hildebrand, PE, 1979b, 'Summary of the Sondeo methodology used by ICTA', ICTA, Guatemala

Hildebrand, PE, 1981, 'Combining disciplines in rapid appraisal: the Sondeo approach', Agricultural Administration, 8, pp 423–432

Hildebrand, PE, 1982, 'Discussion of farming systems research: issues in research strategy and technology design', American Journal of Agricultural Economics, December 1982, pp 905–6

Hill, P, 1972, Rural Hausa: a village and a setting, Cambridge University Press, London

Hoek, A, van den, 1983, 'Landscape planning and design for a watershed in the Kathama Agroforestry Project, Kenya', MSc paper, Department of Landscape Architecture and Planning, Wageningen Agricultural University, Wageningen

Hope, A and Timmel, S, 1984, Training for Transformation (Books 1, 2, 3), Mambo Press, Zimbabwe

Horton, DE, 1984, Social Scientists in Agricultural Research: Lessons from the Montaro Valley Project, Peru, IDRC, Ottawa, 1984

Horton, D and Prain, D, 1987, 'CIP's experience with farmer participation in on-farm research', Proceedings, Taller para America Latina sobre Investigacion de Frijol en Campos de Agricultores, CIAT, Cali, Colombia, 16–25 February

*Hossain, SMA and Islam, MT, 1985, A report on the first workshop of innovative rice farmers, GTI Publication No 56, Bangladesh Agricultural University, Mymensingh

Hossain, SMA, Sattar, M, Ahmed, JV, Salim, M, Islam, MS and Salam, NV, 1987,

205

'Cropping Systems research and farmers' innovativeness in a farming community in Bangladesh', IDS Workshop

Howes, M and Chambers, R, 1979, 'Indigenous technical knowledge: analysis, implications and issues', *IDS Bulletin*, 10 (2), pp 5–11

Huxley, PA, and Wood, PJ, 1984, 'Technology and research considerations in ICRAF's diagnosis and design procedures', ICRAF, Nairobi

ICRAF, 1983a, 'Guidelines for agroforestry diagnosis and design', *ICRAF Working Paper No 6*, ICRAF, Nairobi

ICRAF, 1983b, 'Resources for agroforestry diagnosis and design', *ICRAF Working Paper No 7*, ICRAF, Nairobi

IDS, 1979, *IDS Bulletin*, 10 (2), 'Whose Knowledge Counts?'

ILEIA, 1988a, *Towards Sustainable Agriculture, Part One: Abstracts, Periodicals, Organizations, Part Two: Bibliography*, ILEIA Newsletter, May

ILEIA, 1988b, *Participative Technology Development*, ILEIA Newsletter Vol 4, No 3, October (also in French)

ILEIA, 1989, Proceedings of the ILEIA Workshop on Operational Approaches for Participatory Technology Development in Sustainable Agriculture, ILEIA, Leusden, Netherlands

ILO, 1981, *Zambia: Basic Needs in an Economy Under Pressure*, International Labour Office, Jobs and Skills Programme for Africa, Addis Ababa

ISNAR, 1986, 'The organizational and managerial implications of on-farm research: brief project statement', International Service for National Agricultural Research, The Hague

Jaiyebo, E and Moore, A, 1965, 'Soil fertility and nutrient storage in different soil-vegetation systems in a tropical rainforest environment', *Tropical Agriculture* (Trinidad), 41, pp 129–139

Jama, B, 1986, 'Alley cropping maize (*Zea mays*) and green gram (*Phaseolus aureus*) with *Leucaena leucocephala* as an alternative farming system at Mtwapa, Coast Province, Kenya', MSc thesis, University of Nairobi

*Jama, B, 1987, 'Learning from the farmer', *IDS Workshop*

Jiggins, J, 1988, 'Farmer participatory research and technology development', *Occasional Papers in Rural Extension, No 5*, Department of Rural Extension Studies, University of Guelph, Ontario, Canada

Jiggins, J, Engel, P, and Lightfoot, C, 1988, 'Matrices on different steps of participative technology development'. Workshop on Operational Approaches for Participative Technology Development in Sustainable Agriculture, ILEIA, Leusden, 11–12 April

Johnson, AW, 1972, 'Individuality and experimentation in traditional agriculture', *Human Ecology*, 1 (2), pp 448–459

Juma, C, 1987a 'Genetic resource and biotechnology in Kenya: towards long-term food security', Public Law Institute, Nairobi

*Juma, C, 1987b, 'Ecological complexity and agricultural innovation: the use of indigenous genetic resources in Bungoma, Kenya', IDS Workshop

Kabutha, C and Ford, R, 1988, 'Using RRA to formulate a village resources management plan, Mbusanyi, Kenya', *RRA Notes*, 2, IIED, London

Kang, BT, Wilson, GF and Sipkens, L, 1981, 'Alley cropping with maize (*Zea mays* L.) and leucaena (*Leucaena Leucocephala* LAM) in southern Nigeria', *Plant and Soil*, 63, pp 165–179

Kang, BT, Wilson, GF, and Lawson, TT, 1986, 'Alley cropping as a stable alternative to shifting cultivation', International Institute for Tropical Agriculture (IITA), Ibadan, Nigeria

*Kean, S, 1988, 'Developing a partnership between farmers and scientists: the example of Zambia's Adaptive Research Planning Team', *Experimental Agriculture*, Vol 24, part 3, pp 289–299

Kean, SA and Singogo, Lingston P, 1988, *Zambia: Organization and Management of the Adaptive Research Planning Team (ARPT), Research Branch, Ministry of Agriculture and Water Development*, OFCOR Case Study No 1, ISNAR, The Netherlands, May

Kishewitsch, S, 1987, 'Agroforestry adaptation and adoption in Coast Province, Kenya', MSc Thesis, Faculty of Environmental Studies, York University, Toronto, Canada

KKU, 1987, *Proceedings of the 1985 International Conference on Rapid Rural Appraisal*, Rural Systems Research and Farming Systems Research Projects, Khon Kaen University, Khon Kaen, Thailand

Knipscheer, see Baker and Knipscheer

Knipscheer, HC and Kedi Suradisastra, 1986, 'Farmer Participation in Indonesian Livestock Farming Systems by Regular Research Field Hearings (RRFH)', *Agricultural Administration*, vol 22, pp 205–216

Korten, DC, 1980, 'Community organisation and rural development: a learning process approach', *Public Administration Review*, Vol 40, September-October pp 480–510

Korten, DC, 1984, 'Rural Development Programming: the learning process approach', in Korten, DC and Klauss, R (eds), *People-centered Development: contributions towards theory and planning frameworks*, Kumarian Press, West Hartford

Krishnamoorthy, CH, 1975, 'Pilot development project and operation research project', a mimeograph paper, All-India Coordinated Research Project for Dryland Agriculture, Santoshnagar, Hyderabad 500659, India

Kumar, K, 1987, *Conducting Group Interviews in Developing Countries*, AID Program Design and Evaluation Methodology Report No 8, USAID, Washington

Kuyper, JBH, 1987, 'On-farm agroforestry research in Kisii, Kenya', Beijer Institute, Nairobi

*Lamug, C, 1987, 'Interaction of upland farmers and scientists', IDS Workshop. See section 2.4 for a shortened version

Last, M and Chavunduka, GL (eds), 1986, *The Professionalisation of African Medicine*, Manchester University Press

Lightfoot, C, 1986, 'A short methodological account of a dynamic systems field experiment: the case of legume enriched fallows for the restoration of soil fertility, eradication of Imperata, improvement of pasture, and reduction in labour for cultivation, in the Philippines'. Draft paper for Farming Systems Symposium, Kansas State University, October 5–8, Farming Systems Development Project – Eastern Visayas, Ministry of Agriculture, Tacloban, Leyte, Philippines

Lightfoot, C, 1986, 'Conducting on farm research in FSR: making a good idea work'. Farming Systems Support Project, Gainesville, Florida

Lightfoot, C, 1987, 'Indigenous Research and on-farm Trials', *Agricultural Administration and Extension*, vol 24, pp 79–89

Lightfoot, C, de Guia, O Jr and Ocado, F, 1988, 'A participatory method for systems-problem research rehabilitating marginal uplands in the Philippines', *Experimental Agriculture* vol 24, part 3, pp 301–309

Lightfoot, C, Quero, F Jr and Villanueva, MR, 1985, *Review of research methods and findings, FSDP-EV Project Report No 33*, Department of Agriculture, Tacloban, Philippines

*Lightfoot, C, de Guia, O Jr, Aliman, A and Ocado, F, 1987, 'Letting farmers decide in on-farm research', IDS Workshop. See section 2.7 for a shortened version

Lindsay, RS and Hepper, F, 1978, *Medicinal Plants of the Marakwet*, London, Kew

Long, N, 1986, 'Creating space for change: a perspective on the sociology of development', Wageningen Agricultural University

Lundgren, B, 1982, 'Introduction', *Agroforestry Systems*, 1(1), pp 3–6

McCracken, JA, 1988, *Participatory Rapid Appraisal in Gujarat: a trial model for the Aga Khan Rural Support Programme (Kenya)*, International Institute for Environment and Development, London, November

McCracken, JA, Pretty, JN and Conway, G, 1988, *An Introduction to Rapid Rural Appraisal for Agricultural Development*, International Institute for Environment and Development, London

Makarim, AK and Cassel, DK, 1985, 'Tillage and soil amendments for land reclamation of a bulldozed area in Sitiung, West Sumatra', *Field Research Brief, No 4*, Tropsoils Project, Sitiung, Indonesia

Malaret, L, and Nguru, FN, 1986, 'An ethno-ecological study of the interaction between termites and small-holder farmers in Kenya', abstract of paper presented at the International Conference on Tropical Entomology, 31 August–5 September, ICIPE, Nairobi

Martin, A and Farrington, J, 1987, 'Abstracts of recent field experience with farmer participatory research', *Agricultural Administration (Research and Extension) Network Paper* 22, June, ODI, London

Matlon, P, 1982, 'On-farm experimentation: ICRISAT farmers' tests in the context of a program of farm-level baseline studies', ICRISAT, Ouagadougou, Burkina Faso

Matlon, P, 1985, 'A critical review of objectives, methods and progress to date in sorghum and millet improvement: a case study of ICRISAT/Burkina Faso'. In Ohm, HW and Nagy, J, (eds) *Appropriate Technologies for Farmers in Semi-arid West Africa*, Purdue University, West Lafayette, Indiana, pp 154–179

Matlon, P, and Spencer, DS, 1984, 'Increasing food production in sub-Saharan Africa: environmental problems and inadequate technological solutions', *American Journal of Agricultural Economics*, Vol 66, No 5, pp 671–676

Matlon, P, Cantrell, R, King, D and Benoit-Cattin, M, (eds), 1984, *Farmers' Participation in the Development of Technology: Coming Full Circle*, IDRC, Box 8500, Ottawa, 179 pp

Mathema, SB and Van Der Veen, MG, 1978, 'Socio-economic research on farming systems in Nepal', *Cropping Systems Program, Technical Report 01*, Department of Agriculture, Khumaltar, Nepal

Mathema, SB, Galt, DL, Krishna, KC, Shrestha, RB, Sharma, AR, Upraity, VN and Vaidya, NL, 1986, 'Group survey and on-farm trial process, Naldung Village Panchayat, *SERED Report No 2*, Ministry of Agriculture, Khumaltar, Nepal

*Mathema, SB and Galt, D, 1987, 'The Samuhik Bhraman process in Nepal: a multidisciplinary group activity to approach farmers', IDS Workshop. See section 2.3 above for a shortened version

*Maurya, DM and Bottrall, A, 1987, 'Innovative approach of farmers for raising their farm productivity', IDS Workshop. See section 1.2 for an extract. For a fuller version see Maurya, Bottrall and Farrington 1988

Maurya, DM, Bottrall, A and Farrington, J, 1988, 'Improved livelihoods, genetic diversity and farmer participation: a strategy for rice-breeding in rainfed areas of India', *Experimental Agriculture*, vol 24, part 3, pp 311–320

Maxwell, S, 1984, 'The social scientist in farming systems research', *IDS Discussion Paper* 199, Institute of Development Studies, University of Sussex, November

Maxwell, S, 1986, 'Farming Systems Research: hitting a moving target', *World Development*, vol 14, no 1

Menz, K, 1980, 'Unit farms and farming systems research: the IITA experience', *Agricultural Systems*, 6, pp 45–51

Mercado, B, 1986, 'Control of *Imperata cylindrica*', in Moody, K. (ed), *Weed Control in Tropical Crops Volume 2*, SEARCA, Los Banos, Philippines, 293 pp

Merrill-Sands, D, 1988, 'International Service for National Agricultural Research: Study on the Organization and Management of On-farm Client-oriented Research (OFCOR): Part I: Introduction', *ODI Discussion Paper* 28, Agricultural Administration (Research and Extension) Network, Overseas Development Institute, Regent's College, Regent's Park, London, May

Mozans, HJ, 1983, *Woman in Science*, reprint of 1974, MIT Press, Cambridge, Massachusetts

Munyao, PM, 1987, 'The importance of gathered food and medicinal plant species in Kakuyuni and Kathama areas of Machakos', Annex I in Wachira, KK (ed) *Women's Use of Off-farm and Boundary Lands: agroforestry potentials*, Final Report, ICRAF, Nairobi, pp 56–60

Mutiso, RM, 1987, 'Survey of gathered wild foods and their nutritional value in Kathama', Annex I in Wachira, KK (ed) *Women's Use of Off-farm and Boundary Lands: agroforestry potentials*, Final Report, ICRAF, Nairobi, pp 114–123

Muturi, SN, 1981, 'Agricultural research at the coast', a report of the National Council for Science and Technology, Nairobi, Kenya

Nair, K, 1979, *In Defence of the Irrational Peasant: Indian Agriculture after the Green Revolution*, Chicago University Press

Ngambeki, DS and Wilson, GF (undated), 'Economic and on-farm evaluation of alley cropping with *Leucaena leucocephala*', International Institute for Tropical Agriculture, Ibadan, Nigeria

Norman, DW, 1974, 'Rationalizing mixed cropping under indigenous conditions: the example of northern Nigeria', *Journal of Development Studies*, 11, pp 3–21

Norman, DW, 1980, 'Farming systems approach: relevance for the small farmer', *Rural Development Paper No 5*, Michigan State University, East Lansing

Norman, DW, Simmons, EB and Hays, HM, 1982, *Farming Systems in the Nigerian Savanna: research and strategies for development*, Westview Press, Boulder, Colorado

Norman, DW and Collinson, M, 1985, 'Farming systems approach to research in theory and practice', ACIAR Workshop on Farming Systems Research, Sydney, Australia, 12–15 May, 1985

*Norman, DW, Baker, D, Heinrich, G, Jonas, G, Maskiara, S and Worman, F, 1987, 'Farmer groups for technology development: experiences from Botswana', IDS Workshop. For a shortened version of this paper see section 3.6. For a full version see Norman et al 1988

Norman, D, Baker, D, Heinrich, G and Worman, F, 1988, 'Technology development and farmer groups: experiences from Botswana', *Experimental Agriculture*, Vol 24, part 3, pp 321–331

Oakley, P, 1987, 'State or process, means or end? The concept of participation in rural development', *RRDC Bulletin*, Reading University, March, pp 3–9

Odum, TO, 1983, *Systems Ecology: an introduction*, John Wiley & Sons, New York

Oduol, PA, 1986, 'The shamba system: an indigenous system of food production from forest areas in Kenya', *Agroforestry Systems*, 4, pp 365–373

Okali, C, 1983, *Cocoa and Kinship in Ghana: the matrilineal Akar of Ghana*, Routledge and Kegan Paul, London

Okali, C and Milligan, K, 1982, article in 'The Role of Social Scientists in Developing Food Production Technology', IRRI, Manila, Philippines

Okali, C and Knipscheer, H, 1985, 'Small ruminant production in mixed farming systems: case studies in research design', paper prepared for 5th Annual FSSP Research and Extension Symposium, Kansas State University

Okali, C and Sumberg, JE, 1986a, 'Examining divergent strategies in farming systems research', *Agricultural Administration*, 22, pp 233–253

Okali, C and Sumberg, JE, 1986b, 'Sheep and goats, men and women: household relations and small ruminant development in south-west Nigeria', *Agricultural Systems*, 18, pp 39–59

Pacey, A and Cullis, A, 1986, *Rainwater Harvesting*, Intermediate Technology Publications, London

Passerini, E, 1986, 'Food for everyone? Yes, from trees', *Agriculture and Human Values*, 3, (3), pp 15–20

Pope, E, 1986, 'Importance of indigenous wild foods for women in the Kathama area, Machakos District, Kenya', paper presented in the KENGO seminar on the role of indigenous plants in our lives, 15 July

Posey, 1984, 'Ethnoecology as applied anthropology in Amazonian development', *Human Organization* 43(2), pp 95–107

Pretty, JN, ed, 1988, *Alpuri: Rapid Agroecosystem Zoning*, Malakand Fruit and Vegetable Development Project, Mingora, and International Institute for Environment and Development, London

Raintree, JB and Young, A, 1983, 'Guidelines for agroforestry diagnosis and design', International Council for Research in Agroforestry, Nairobi

*Raman, KV, 1987, 'Scientists' training: experiences in promoting interaction with the farmers', IDS Workshop. See section 4.2 for a shortened version

Reed, C, 1977, *Origins of Agriculture*, Mouton Publishers, The Hague

*Repulda, RT, Quero, F, Ayaso, R, Guia, O de and Lightfoot, C, 1987, 'Doing research with resource poor farmers: FSDP-EV perspectives and programmes', IDS Workshop

Rhoades, RE, 1982, *The Art of the Informal Agricultural Survey*, International Potato Centre, Lima, Peru

Rhoades, RE, 1983, 'Tecnicista versus campesinista: praxis and theory of farmer involvement in agricultural research. A post-harvest example from the Andes'. Paper presented at a workshop on Farmers' Participation in the Development and Evaluation of Agricultural Technology, ICRISAT/SAFGRAD/IRAT, Ouagadougou, 20–24 September

Rhoades, RE, 1984, *Breaking New Ground: Agricultural Anthropology*, International Potato Centre, Lima, 71 pp

*Rhoades, RE, 1987, 'The role of farmers in the creation and continuing development of agri-technology and systems', IDS Workshop. See section 1.1 for a shortened version. Also available as 'Farmers and Experimentation', ODI Agricultural Administration (Research and Extension) Network *Discussion Paper* 21, December

Rhoades, RE and Booth, RH, 1982, 'Farmer-back-to-farmer: a model for generating acceptable agricultural technology', *Agricultural Administration*, Vol 11, pp 127–137

Rhoades, RE, Booth, RH, Shaw, R and Werge, R, 1985, 'The role of anthropologists in developing improved technologies', *Appropriate Technology*, Vol 11, No 4, pp 11–13

Rhoades, RE, Horton, DE and Booth, RH, 1986, 'Anthropologist, biological scientist and economist: the three musketeers, or three stooges of farming systems research?' in JR Jones and BJ Wallace (eds), *Social Sciences and Farming Systems Research*, Westview Press, Boulder, Colorado

Rhoades, RE and Bidegaray, P, 1987, 'The farmers of Yurimaguas', CIP (International Potato Centre), Lima, Peru

Rhoades, RE and Bebbington, A, 1988, 'Farmers who experiment: an untapped resource for agricultural research and development', paper presented at the International Congress on Plant Physiology, New Delhi, 15–20 February

Richards, P, 1985, *Indigenous Agricultural Revolution*, Hutchinson, London and Westview Press, Boulder, Colorado

*Richards, P, 1987, 'Agriculture as a performance', IDS Workshop. See section 1.6 above for a shortened version

*Rocheleau, D, 1987a, 'The user perspective and the agroforestry research and action agenda', IDS Workshop. Published as Chapter 6 in H. Gholz (ed), *Agroforestry*, Martinus Nijhoff, Dordrecht, Netherlands, September 1987

Rocheleau, D, 1987b, 'Women, trees and tenure: implications for agroforestry research'. In Raintree, JB (ed), *Trees and Tenure: Proceedings of an International Workshop on Tenure Issus in Agroforestry*, International Council for Research in Agroforestry, Nairobi, and Land Tenure Centre, Madison, Wisconsin

*Rocheleau, D, 1987c, 'Floppy disc – paper of profound importance lost in the memory of an agroforestry Apple – not seen by the editor', IDS Workshop

Rocheleau, DE, Khasiala, P, Munyao, M, Mutiso, M, Opala, E, Wanjohi, B, and Wanjuagna, A, 1985, 'Women's use of off-farm lands: implications for agroforestry research', project report to the Ford Foundation, ICRAF, Nairobi, mimeo

Rocheleau, D and Weber, F, 1987, 'Agroforestry for soil and water conservation in dryland Africa', (draft handbook for community-based research in agroforestry), ICRAF, Nairobi

Ruano, Sergio A et al, 1982, 'Tecnicas Basicas de Entrevista al Realizar Investigacion sobre sistemas de cultivos', ICTA, Guatemala

Sagar, D, and Farrington, J, 1988, 'Participatory approaches to technology generation: from the development of methodology to wider-scale implementation', *Agricultural Administration (Research and Development) Network Paper No 2*, ODI, London, December

Sajise, P, 1981, 'Experimental education and its transfer: the Philippine experience',

210

paper presented at the UNESCO-Rihed conference, Education in ASEAN Universities, Malaysia, 17–21 August

Sajise, P, 1984, 'Plant succession and agrosystem management', in Rambo, AT and Sajise, P (eds), *An Introduction to Human Ecology Research on Agricultural Systems in Southeast Asia*, UP Los Banos, Philippines, 327 pp

*Sanghi, NK, 1987, 'Participation of farmers as co-research workers: some case studies in dryland agriculture', IDS Workshop. See section 4.3 for a shortened version. Also available as an ODI Network Paper

Sangi, NK, Vishnu Murthy, T, Kameshwara Rao, V, Prabhanjana Rao, SB, Vijayalakshmi, K, Venkateswarlu, J, Gobindswamy, S and Atwal, JS, 1983, 'Operation research in dryland agriculture for semi-arid red soils of Hyderabad', Project Bulletin No 3, All-India Coordinated Research Project for Dryland Agriculture, Hyderabad 500 659

Schlippe, P de, 1947, 'Une methode cultural conservatrice-adaptation empirique indigine', *Semaine agricole de Yangambi*, INEAC

Schlippe, P de, 1953, 'Le revoluement rural en fonction de notre connaissance de la coutume agricole', INCIDI, The Hague

Schlippe, P de, 1956, *Shifting Cultivation in Africa: the Zande System of Agriculture*, Routledge and Kegan Paul, London

Schlippe, P de, 1957, 'Methodes de recherches quantitative dans l'economie rurale coutumiere de l'Afrique centrale', Directeur de l'Agriculture de Forets et de l'Elevage, Bruxelles

Shaner, WW, Philipp, PF and Schmehl, WR, 1982, *Farming Systems Research and Development: Guidelines for Developing Countries*, Westview Press, Boulder, Colorado

Shrestha, RB, Sharma, AR and Galt, DL, 1987, 'Baseline Survey Report of Baglung District, and Naldung Key Informant Survey Report, *SERED Reports Nos 4 and 6*, Ministry of Agriculture, Socio-Economic Research and Extension Division, Khumaltar, Nepal

*Smutkupt, S, 1987, 'Farmers to farmers: researchers' role as facilitators', IDS Workshop

Spedding, CRW 1979, *An Introduction to Agricultural Systems*, Applied Science Publishers, London, 169 pp

Stone, RM, 1982, *Let the inside be sweet: the interpretation of music events among the Kpelle of Liberia*, Bloomington: Indiana University Press

Stoop, WA, 1988, *NARS Linkages in Technology Generation and Technology Transfer*, ISNAR Working Paper No 11, International Service for National Agricultural Research, The Hague, April

Sumberg, JE and Okali, C, 1984, 'Linking crops and animal production: pilot development programmes for smallholders in southwest Nigeria', *Rural Development in Nigeria*, 1, pp 25–9

*Sumberg, JE and Okali, C, 1987, 'Farmers, on-farm research and the development of new technology', IDS Workshop. See section 3.1 for a shortened version. See Sumberg and Okali 1988 for a fuller version

Sumberg, J and Okali, C, 1988, 'Farmers, on-farm research and the development of new technology', *Experimental Agriculture*, Vol 24, part 3, pp 333–342

Sutherland, A, 1986, 'Extension workers, small-scale farmers, and agricultural research: a case study in Kabwe Rural, Central Province, Zambia', *Agricultural Administration (Research and Extension) Network Paper* No 15 ODI, London, March

Swift, J, 1979, 'Notes on traditional knowledge, modern knowledge, and rural development'. *IDS Bulletin* Vol 10 (2), pp 41–43

Swindale, LD, 1987, 'Farming systems and the International Agricultural Research Centres: an interpretative summary', in Proceedings of the Workshop on Farming Systems Research, ICRISAT, Patancheru, Andhra Pradesh, India

Thiele, G, Davis, P, and Farrington, J, 1988, 'Strength in diversity: innovation in agricultural technology development in eastern Bolivia', *Agricultural Administration (Research and Extension) Network Paper No* 1, ODI, London, December

*Thrupp, LA, 1987, 'Building legitimacy of indigenous knowledge: empowerment for third world people, or 'scientized packages' to be sold by development agencies', IDS Workshop

Torres, F, 'Networking for the generation of agroforestry technologies in Africa', ICRAF, Nairobi

Vayda, AP, Colfer, CJP and Brotokusumo, M, 1980, 'Interactions between people and forests in East Kalimantan', in *Impact of Science on Society*, Vol 30 (3), pp 179–190

*Verma, GP, 1987, 'Farmers' participation in watershed management', IDS Workshop

von der Osten, A, Ewell, PT and Merrill-Sands, D, 1988, *Organization and Management of Research for Resource-Poor Farmers*, Staff Notes No. 88–13, ISNAR, The Netherlands, September

Vonk, RB, 1983, 'Report on Methodology and Technology Generating Exercise', Wageningen Agricultural University, Wageningen

Vonk, RB, 1986, 'Report on Siaya Agroforestry Project', CARE-Kenya, Nairobi

Wachira, KK, 1987, 'Women's use of off-farm and boundary lands: agroforestry potentials', ICRAF, Nairobi

Wade, MK, Fahmuddin Agus, D, and Colfer, CJP, 1985, 'The contribution of farmer-managed research in technology development'. International Farming Systems Workshop, Sukarami, West Sumatra

Wanjohi, B, 1987, 'Women's groups gathered plants and their agroforestry potentials in the Kathama area', Annex 1 in Wachira, KK (ed), *Women's Use of Off-farm and Boundary Lands: agroforestry potentials*, Final Report, ICRAF, Nairobi, pp 61–104

Waters-Bayer, A, 1989, 'Trails by scientists and farmers: opportunities for cooperation in ecofarming research', in Kotschi, J, (ed) *Ecofarming Practices for Tropical Smallholdings – Research and Development in Technical Cooperation*, GTZ, Eschborn pp 161–183

Waters-Bayer, A, and Bayer, W, 1988, 'Zero-station livestock systems research: pastoralist-scientist cooperation in technology development', paper for workshop on Operational Approaches for Participative Technology Development in Sustainable Agriculture, ILEIA, Leusden, 11–12 April

Watts, M, 1983, *Silent Violence: food, famine and peasantry in northern Nigeria*, Berkeley: University of California Press

Whyte, WF, 1981, *Participatory Approaches to Agricultural Research and Development: a state-of-the-art paper*, Rural Development Committee, Center for International Studies, Cornell University, Ithaca, NY

Wilson, KB, 1987, 'Research on trees in the Mazvihwa and surrounding areas', unpublished report for ENDA-Zimbabwe, Harare

Worman, FD, and Heinrich, GM, 1988, 'Two operational approaches to participative technology development used by the Agricultural Technology Improvement Project, Francistown, Botswana'. Workshop on Operational Approaches for Participative Technology Development in Sustainable Agriculture, ILEIA, Leusden, 11–12 April

Wright, P, 1985, 'Water and soil conservation by farmers', in Ohm, HW and Nagy, JG, *Applied Technology for Farmers in Semi-arid Africa*, Purdue University, West Lafayette, International Programs in Agriculture, pp 56–60

Wright, P and Bonkougou, EC, 1986, 'Soil and water conservation as a starting point for rural forestry: the Oxfam project in Ouahigouya, Burkina Faso', *Rural Africana*, 23–24, pp 79–85

Zandstra, HG, Price, EC, Litsinger, JA and Morris, RA, 1981, *A Methodology for On-farm Cropping Systems Research*, International Rice Research Institute, Los Banos, Philippines

Index

adaptation of technology 5–7, 49, 178, 179, 185
adoption of technology 114, 158, 180, 189, 192
aerial photographs 55, 89
AF *see* agroforestry
Africa 14–16, 38, 42, 49, 192
  West 39, 58, 87, 110, 113, 185
AICRPDA *see* All India Coordinated Research Project on Dryland Agriculture
Aga Khan Rural Support Programme (AKRSP) 84, 192
Agricultural Research and Production Project (ARPP) 68, 191
Agricultural Research Service (India) (ARS) 171
Agricultural Technology Improvement Project (ATIP) 136–9, 142–6, 191
agriculture as performance 39–43, 185
agroecology 16, 173
agroforestry (AF) 14–24, 74, 110, 112, 155, 156, 157
agronomists 42, 63–4, 92, 121, 127, 130, 151
agronomy 16, 62, 92, 104
All India Coordinated Research Project on Dryland Agriculture (AICRPDA) 24, 176
alley cropping 18, 46, 110, 112–14, 185
Andhra Pradesh Agricultural University 170
anthropologists 62, 63, 102, 127, 151
anthropology 16, 155
ARPP *see* Agricultural Research and Production Project
ARS *see* Agricultural Research Service
ATIP *see* Agricultural Technology Improvement Project
Bangladesh 37, 87, 90, 91, 102, 123
  cropping patterns 34–6, 185
  innovator workshops 44, 132–6, 186
  unusual practices 26–8
Bangladesh Agricultural Research Institute (BARI) 26, 28, 87, 133, 135

Bangladesh Agricultural University (BAU) 134
BARI *see* Bangladesh Agricultural Research Institute
BAU *see* Bangladesh Agricultural University
beans 57, 117–18, 120–1, 125, 127, 134
BFD *see* Bureau of Forest Development
biographical analysis 62, 64, 67, 105, 183, 185
biological scientists 9, 24, 28, 30, 69
Bolivia 192
Botswana 32, 136–46, 191
'bottom up' approach 46, 136, 187, 188
Brazil 109, 123
Bureau of Forest Development (BFD) 73, 74, 75, 77
bureaucracy 63, 64, 168, 181–2, 186–8, 192–4
Burkina Faso 49
bush beans 109, 115–17, 119, 128–30, 131

cassava 47, 58, 61–7, 99, 109, 113, 115–19, 125–8, 130, 152
Central America 34, 36, 56
Central Mennonite Committee 192
cereals 48, 60, 149, 181
  see also maize; wheat
CGIAR *see* Consultative Group for International Agricultural Research
changes 36, 56, 65, 89, 185
  institutional 161, 169, 175–95
chillies 132, 134, 154
choice of technologies 10, 47, 183, 184
CIAT *see* International Centre for Tropical Agriculature
CIMMYT *see* International Centre for the Improvement of Corn and Wheat
CIP *see* International Potato Centre
classifications 31–2, 65, 156
coconut 155, 156
coffee 34, 36, 156

213

217

upland farmers, (Philippines) 73–7, 160
USAID *see* United States Agency for
International Development

varietal selection *see* crop varieties
vegetables 21–3, 28–9, 58, 99, 132, 149,
152, 154
Virgilio 61
visual aids 44, 183

water 29, 30, 84, 88
conservation 49, 178
drought 40, 81, 84, 90, 137, 159, 181
flooding 90, 134
irrigation 8, 9, 34, 35, 133
rainfall 40, 81, 137, 151, 159
watermelon 133, 134
weeds 12, 130, 148
cogon grass 93–100, 103, 160
wheat 12, 34, 72, 132, 133, 134
Winrock International Institute for
Agricultural Development 68
witchcraft 42

WN *see* World Neighbors
Wodaabe Fulani people (Niger) 87, 88
women 15, 81, 116, 124, 129, 131, 138,
154
domestication of crops 4, 22, 47
farmers 18–19, 65, 69, 70, 101, 146
heading households 125, 139, 152
land inheritance 156
map drawing 87, 88
responsibilities 20–2, 29, 45, 47, 91,
148
scientists/researchers 29, 30,
101–2
World Neighbors (WN) programmes
56, 57, 58, 192

yams 34, 113
yields 9, 110, 114, 133, 137
increased 36, 39, 130

Zaire 65, 103
Zambia 31–2, 47, 89, 92, 125–6, 187,
191